Stories of Dangerous Weapons

Kumar Aansu

Published by A2S, 2026.

STORIES OF DANGEROUS WEAPONS

First edition. March 25, 2026.

ISBN: 979-8233885877

Written by Kumar Aansu.

Table of Contents

Preface

This book presents the stories of many types of weapons that exist in the world today and explains how they affect human life. Many people hear about weapons in news reports, films, or conversations, but often the information is unclear, confusing, or incomplete. Some descriptions are too technical, while others are too dramatic and not based on facts. This book tries to take a balanced path. It explains weapons clearly and simply so that anyone, including young readers, can understand what they are, how they work, and why they matter.

The idea behind this book came from a common problem. People often know the names of weapons but do not fully understand them. For example, someone may have heard of a nuclear bomb, a drone, or a missile, but may not know how it actually works or what impact it has on people and society. This lack of understanding can lead to fear, confusion, or even wrong beliefs. By providing clear and honest explanations, this book aims to replace confusion with knowledge.

Another important reason for writing this book is to show that weapons are not just objects. They are connected to human decisions, human history, and human values. Every weapon described in this book has a story. Some were created during times of war. Others were developed as part of scientific progress. Some were meant to protect, while others were designed to attack. By learning about these weapons, readers can also learn about the choices people have made and the consequences of those choices.

The book also helps readers understand that not all weapons cause damage in the same way. Some cause physical destruction, like bombs or artillery. Others, such as cyber tools or psychological methods, affect systems, information, or human thinking. This shows that harm is not always visible. Sometimes, the most serious effects are not easily seen, such as fear, confusion, or disruption of daily life. By including a wide range of examples, this book gives a complete picture of how different types of weapons operate.

The purpose of this book is not to promote weapons or to create fear. Instead, it is to inform and educate. When people understand how things work, they are better able to think clearly and make responsible decisions. Knowledge helps people ask questions, understand risks, and value peace. This book encourages readers to think about the impact of weapons, not just their power. It also helps readers see how important it is to use knowledge responsibly.

In this book, each topic is explained in a simple, direct way. The text avoids complex words and difficult ideas so that readers do not feel lost or overwhelmed. Each section describes what the weapon is, how it was developed, and how it works in general terms. It also explains the advantages and disadvantages, as well as the effects on people, communities, and the environment. The goal is to provide a clear and complete understanding without going into unnecessary detail.

Readers can use this book in different ways. It can be read from beginning to end, like a story that explores different types of weapons one by one. It can also be used as a reference, where readers choose a topic they are curious about and learn about it. Students may find it helpful for school projects. Teachers can use it to explain complex topics. Even casual readers can use it to improve their general knowledge about the world.

One important thing to remember while reading this book is that knowledge brings responsibility. Understanding weapons should not lead to fear or misuse. Instead, it should encourage thoughtful thinking and respect for human life. The book gently reminds readers that while technology can be powerful, it must be guided by wisdom and care. The future depends on how people choose to use knowledge.

Readers should read this book because it offers clarity in a world where information is often confusing. It presents facts simply and honestly, without exaggeration or unnecessary detail. It helps readers understand not only how weapons work, but also why they matter. This understanding can lead to better awareness of global events and a deeper appreciation for peace and stability.

Another reason to read this book is that it connects science with real life. Many of the ideas behind weapons come from basic science, such as energy, motion, and communication. By learning about these ideas in a simple context, readers can also improve their understanding of science. This makes the book both educational and practical.

This book is also a reminder that human actions shape the world. People create weapons, and people decide how they are used. By learning about them, readers can better understand the importance of responsibility, cooperation, and careful decision-making. It shows that while technology can be powerful, it is human values that determine how it is used.

In the end, this book is about understanding. It brings together information on many types of weapons and presents it clearly and meaningfully. It invites readers to learn, think, and reflect. It does not try to give all the answers, but it provides a strong foundation for asking the right questions. Through this knowledge, readers can become more aware of the world around them and the role they play in shaping it.

—Author

1. Nuclear Bomb

A nuclear bomb is one of the most powerful and destructive weapons ever created by humans. It releases a huge amount of energy from atomic nuclei, either through nuclear fission, where heavy atoms split into smaller parts, or through nuclear fusion, where light atoms combine to form heavier ones. This release of energy happens in a fraction of a second, but its effects can last for many years. The idea of using atomic energy for weapons became real during the Second World War, when scientists discovered how to control and trigger nuclear reactions. The first nuclear bombs were developed under a secret research effort in the United States known as the Manhattan Project. In 1945, two atomic bombs were dropped on the Japanese cities of Hiroshima and Nagasaki, marking the first and only time nuclear weapons have been used in war. These events changed the course of history and showed the world the extreme power of such weapons.

The basic working principle of a nuclear bomb depends on creating a rapid and uncontrolled nuclear reaction. In simple terms, special materials are used that can release energy when their atoms are split or fused. When this reaction starts, it quickly grows into a chain reaction, releasing enormous heat, light, and pressure. While the scientific ideas behind this process are well known, the actual design, engineering, and handling of such weapons are extremely complex and tightly controlled by governments. It involves advanced technology, strict safety measures, and highly specialized knowledge. Because of the danger involved, detailed methods of building such weapons are not openly shared and are restricted under international laws and agreements.

The historical background of nuclear weapons is closely linked to scientific discoveries in physics during the early twentieth century. Scientists learned that atoms were not solid and unchangeable, but instead had internal parts that could release energy. During the 1930s, experiments showed that splitting certain atoms could produce large amounts of energy. As World War II intensified, there was fear that such knowledge could be used to create powerful weapons. This led to rapid research and development efforts. After the bombings of Hiroshima and Nagasaki, many countries realized both the power and the danger of nuclear weapons. This started a period known as the nuclear arms race, especially between the United States and the Soviet Union during the Cold War. Over time, several other countries also developed nuclear weapons, leading to global concern about safety and control.

The advantages of nuclear bombs are mainly seen from a military and strategic perspective. They act as a strong deterrent, meaning countries with them may be less likely to be attacked because of the fear of retaliation. This idea is often called "mutual deterrence," in which opposing sides avoid war because both have the power to inflict massive destruction. Some leaders argue that nuclear weapons have helped prevent large-scale wars between powerful nations since World War II. They can also end conflicts very quickly due to their immense destructive power, though this comes at a very high human and moral cost.

However, the disadvantages of nuclear bombs are far greater and more serious. The most immediate effect of a nuclear explosion is massive destruction. Buildings are destroyed, fires spread quickly, and people close to the explosion are killed instantly due to extreme heat and pressure. Beyond the initial blast, nuclear bombs release harmful radiation. This radiation can cause severe injuries, sickness, and death even after the explosion. Survivors may suffer from burns, cancer, and other long-term health problems. Radiation can also affect future generations by causing genetic damage.

The environmental impact of nuclear bombs is also severe. The explosion can damage soil, water, and air, rendering large areas uninhabitable. Plants and animals can die or become affected by radiation. In some cases, land remains contaminated for many years. If multiple nuclear weapons are used, the effects could spread across the globe. Scientists have warned about a possible "nuclear winter," where smoke and dust from explosions block sunlight, leading to a drop in global temperatures and affecting food production. This could cause widespread hunger and suffering far beyond the areas directly hit by bombs.

The impact on human civilization is deeply troubling. The use of nuclear weapons can destroy entire cities within seconds, erasing homes, schools, hospitals, and cultural sites. It disrupts social systems, economies, and governments. Survivors often face long-term hardship, including loss of family, health issues, and lack of basic resources. The psychological effects are also significant, as people experience fear, trauma, and uncertainty. The threat of nuclear war has influenced global politics for decades, shaping international relations and defense policies.

Because of these dangers, many efforts have been made to control and reduce nuclear weapons. International treaties like the Treaty on the Non-Proliferation of Nuclear Weapons (NPT) are designed to inhibit the proliferation of nuclear arms and promote disarmament. Many countries support the idea of reducing nuclear arsenals and eventually eliminating these weapons. However, complete disarmament has been difficult to achieve due to concerns about national security and trust between countries.

Another important point is the cost of nuclear weapons. Developing, maintaining, and securing these weapons requires huge financial resources. Governments spend billions of dollars on nuclear programs, including research, testing, and storage. These funds could otherwise be used for education, healthcare, and development. This raises ethical questions about the use of resources in a world where many people still lack basic needs.

There are also safety risks even when nuclear weapons are not used in war. Accidents, technical failures, or human error could lead to unintended explosions or radiation leaks. There have been several close calls in history where misunderstandings or system errors almost led to nuclear conflict. This shows how dangerous these weapons are, not only in war but also in times of peace.

Despite its destructive nature, nuclear technology itself is not entirely harmful. The same scientific principles are used in nuclear power plants to generate electricity. When used carefully and responsibly, nuclear energy can provide large amounts of power without producing air pollution. This shows that the knowledge behind nuclear reactions can be used for both harmful and beneficial purposes, depending on how it is applied.

In conclusion, a nuclear bomb is a powerful and dangerous weapon that has had a deep impact on human history. It was developed during a time of war and has since shaped global politics and security. While it may serve as a deterrent, its disadvantages far outweigh its benefits. The destruction, suffering, and long-term damage caused by nuclear weapons make them one of the greatest threats to humanity. Understanding their history and effects is important so that people can make informed decisions about their future. The hope shared by many around the world is that these weapons will never be used again and that efforts toward peace and cooperation will continue to grow.

2. Hydrogen Bomb

A hydrogen bomb, also known as a thermonuclear bomb, is one of the most powerful weapons ever created by humans. It works by using nuclear fusion, a process in which light atomic nuclei combine to form heavier nuclei, releasing a very large amount of energy. This is the same basic process that powers the sun and other stars. Compared to earlier nuclear weapons that rely mainly on nuclear fission, hydrogen bombs can produce much greater explosive force. The energy released from a hydrogen bomb can be hundreds or even thousands of times stronger than that of the atomic bombs used during the Second World War. This makes it an extremely destructive weapon capable of causing widespread damage across very large areas.

The development of the hydrogen bomb came after the creation of the first atomic bombs. In 1945, the world saw the effects of nuclear fission weapons when bombs were dropped on Hiroshima and Nagasaki. These events showed both the power and the danger of nuclear weapons. Soon after, scientists and military planners began to explore ways to create even more powerful weapons. During the late 1940s and early 1950s, research into nuclear fusion weapons increased, especially in the United States and the Soviet Union. In 1952, the United States tested the first hydrogen bomb, and the Soviet Union followed shortly after with its own test. This marked the beginning of a new stage in the nuclear arms race, where countries competed to build more powerful and advanced weapons.

The working principle of a hydrogen bomb is more complex than that of a simple atomic bomb. It uses a combination of both nuclear fission and nuclear fusion. A small fission explosion is first used to create the extremely high temperatures and pressures needed to initiate the fusion reaction. Once the fusion process begins, it releases a massive amount of energy in a very short time. This chain of events results in an explosion of enormous power. While the scientific idea behind this process is understood, the actual design and construction of such weapons involve highly advanced technology and strict control. The detailed methods are not publicly shared and are tightly regulated by governments because of the serious risks involved.

The effects of a hydrogen bomb explosion are devastating. The initial blast creates a powerful shockwave that can destroy buildings, roads, and other structures over a wide area. The heat generated is so intense that it can cause fires and severe burns even at great distances from the explosion. In addition to the blast and heat, a hydrogen bomb releases large amounts of radiation. This radiation can harm or kill living beings and can remain in the environment for a long time. The explosion also produces a characteristic mushroom-shaped cloud that rises high into the atmosphere, spreading radioactive materials over large regions.

From a military point of view, hydrogen bombs are considered strategic weapons. Their main advantage lies in their ability to act as a strong deterrent. Countries that possess such weapons may be less likely to be attacked because of the fear of massive retaliation. This concept has played an important role in global politics, especially during the Cold War, when major powers avoided direct conflict, in part because of the threat of nuclear war. Some leaders believe that the presence of such powerful weapons has helped prevent large-scale wars between nations with advanced military capabilities.

However, the disadvantages of hydrogen bombs are far more serious and far-reaching. The most obvious drawback is the scale of destruction they can cause. A single hydrogen bomb can destroy an entire city and cause damage far beyond the immediate area. Unlike conventional weapons, which affect smaller areas, a thermonuclear explosion can wipe out large regions in seconds. The loss of human life would be enormous, and the destruction of homes, infrastructure, and essential services would leave survivors in extremely difficult conditions.

Another major concern is the effect of radiation. People exposed to radiation may suffer from severe health problems, including burns, sickness, and long-term diseases such as cancer. Radiation can also affect future generations by causing genetic damage. This means that the harm caused by a hydrogen bomb is not limited to the moment of the explosion but can continue for many years. The environment also suffers greatly. Soil, water, and air can become contaminated, making areas unsafe for living and farming. Plants and animals can be damaged or killed, leading to long-term ecological imbalance.

The impact on human civilization is profound. The use of hydrogen bombs could lead to the destruction of cultural heritage, economic systems, and social structures. Cities that have taken decades or centuries to build could be erased in a moment. Essential services such as healthcare, transportation, and communication would collapse in affected areas. Survivors would face shortages of food, water, and shelter. The psychological impact would also be severe, as people would experience fear, trauma, and uncertainty about the future.

There is also a global risk associated with hydrogen bombs. If multiple such weapons were used in a conflict, the effects could spread far beyond the countries involved. Scientists have suggested that large-scale nuclear explosions could lead to a situation known as nuclear winter. In this scenario, smoke and dust from explosions would rise into the atmosphere and block sunlight, causing a drop in global temperatures. This could affect agriculture and food supplies worldwide, leading to widespread hunger and suffering. Even countries not directly involved in the conflict could face serious consequences.

The financial cost of developing and maintaining hydrogen bombs is also very high. Governments spend large amounts of money on research, testing, security, and storage of these weapons. This raises important questions about how resources are used, especially when many parts of the world still face challenges such as poverty, lack of education, and limited access to healthcare. The existence of such weapons also requires constant monitoring and control to prevent accidents or misuse.

Safety is another major concern. Even without being used in war, hydrogen bombs pose risks. Accidents, technical failures, or human errors could lead to serious incidents. There have been cases in history where misunderstandings or system malfunctions brought countries close to nuclear conflict. These situations highlight how dangerous these weapons are, even when they are not actively used.

In response to these dangers, the international community has taken steps to limit the spread and use of nuclear weapons, including hydrogen bombs. Agreements and treaties have been created to reduce the number of such weapons and to prevent new countries from developing them. Efforts toward disarmament continue, although progress has been slow due to concerns about national security and trust between nations. Many people and organizations around the world support the goal of reducing and eventually eliminating nuclear weapons.

It is also important to understand that the science behind hydrogen bombs is not inherently harmful. Nuclear fusion, if controlled properly, has the potential to provide a clean and powerful source of energy. Scientists are working on ways to use fusion for peaceful purposes, such as generating electricity. This shows that the same knowledge that led to the creation of destructive weapons can also be used to benefit humanity when applied responsibly.

In conclusion, the hydrogen bomb is a weapon of extreme power that represents both scientific achievement and danger. It was developed during a time of global tension and has played a major role in shaping modern history. While it may serve as a deterrent in military strategy, its destructive capacity and long-term effects make it one of the greatest threats to humanity. The damage it can cause to human life, the environment, and global stability is immense. Understanding its nature and impact is important for societies to make informed decisions about its future. Many people hope that continued efforts toward peace, cooperation, and responsible use of science will reduce the risk posed by such weapons and help build a safer world.

3. Chemical Weapons (Nerve Agents)

These include substances like sarin and VX. They were developed for warfare to disrupt the nervous system. They can be dispersed as gas or liquid. Even a small exposure can cause death quickly. They are dangerous because they are hard to detect and spread rapidly.

Chemical weapons known as nerve agents are among the most harmful substances created for use in war. These chemicals are designed to interfere with the normal working of the human nervous system, which controls breathing, movement, and many other vital functions. When a person is exposed to a nerve agent, even in a very small amount, the body loses its ability to send proper signals between nerves and muscles. This can quickly lead to loss of control over the body, severe distress, and often death if medical help is not given at once. Some well-known examples of nerve agents include sarin and VX, both of which are highly toxic and have been used in conflicts and attacks in the past.

The development of nerve agents began in the early twentieth century, with major progress made during the 1930s and 1940s. Scientists were originally studying chemicals for agricultural use, such as pesticides, but during this research, they discovered compounds that were far more toxic than expected. In Germany, during the period leading up to the Second World War, researchers developed some of the first nerve agents, including tabun and sarin. These substances were created as part of efforts to build stronger chemical weapons. Although large amounts were produced, they were not widely used during the war. After the war ended, other countries began their own research programs, leading to the development of even more dangerous agents, such as VX, during the Cold War. Over time, the world became more aware of the extreme risks posed by these chemicals.

Nerve agents work by blocking an important enzyme in the body that helps control nerve signals. This enzyme normally breaks down a chemical messenger that carries signals between nerves and muscles. When the enzyme is blocked, the signals continue, causing muscles to contract repeatedly. This leads to symptoms such as difficulty breathing, loss of muscle control, convulsions, and eventually failure of vital organs. The speed of these effects depends on the type of agent and the level of exposure, but in many cases, symptoms can begin within seconds or minutes.

These chemicals can be spread in various ways, making them especially dangerous. They may be released into the air as gases or vapors, allowing people to inhale them without knowing. They can also exist as liquids that contaminate surfaces, clothing, or water supplies. Some nerve agents are colorless and have little or no odor, making them very hard to detect without special equipment. This means people may not realize they have been exposed until symptoms begin, which can reduce the chances of survival. Their ability to spread quickly and silently makes them a serious threat in both military and civilian settings.

From a military point of view, one of the reasons nerve agents were developed is their ability to disable or kill large numbers of people without causing major damage to buildings or equipment. This may be seen as an advantage in certain types of warfare, where the goal is to control an area rather than destroy it. They are also relatively easy to transport and can be used in small quantities to achieve significant effects. In some cases, their use has been intended to create fear and confusion among populations, which can weaken resistance and disrupt normal life.

However, the disadvantages and dangers of nerve agents are far greater than any possible military benefit. The most serious issue is the suffering they cause. Victims of nerve agent exposure often experience intense pain, panic, and loss of control over their bodies. Death can occur quickly, but in some cases, people may survive with long-term health problems, including damage to the brain, lungs, and other organs. Medical treatment requires fast action and special drugs, and even then, recovery is not always complete.

The impact on human civilization is deeply negative. The use of nerve agents breaks basic rules of humanity and has been widely condemned around the world. These weapons do not distinguish between soldiers and civilians, meaning that innocent people, including children and older people, are often among the victims. Attacks involving nerve agents can destroy trust within communities and create long-lasting fear. People may feel unsafe in their own homes, and normal daily activities can be disrupted for long periods.

The environmental effects of nerve agents also raise serious concerns. When these chemicals are released, they can contaminate soil, water, and surfaces. Some agents break down quickly, but others can remain dangerous for longer periods, especially in enclosed or protected areas. This can make it unsafe for people to return to affected locations. Cleaning up contaminated areas is difficult and requires trained teams with proper equipment. Wildlife and plants can also be harmed, leading to damage in local ecosystems.

There have been several incidents in modern history where nerve agents were used, which helped the world understand their impact more clearly. These events showed how even small amounts of these chemicals can cause widespread harm and panic. As a result, many countries agreed to limit and ban the use of chemical weapons. One of the most important steps in this direction was the creation of the Chemical Weapons Convention, an international agreement that aims to eliminate chemical weapons and prevent their use. Many nations have joined this agreement and have worked to destroy existing stockpiles.

Another important point is the risk of misuse. Because nerve agents can be very effective in small amounts, there is concern that groups or individuals could use them outside of formal military control. This has led to strict regulations on the production, storage, and handling of such chemicals. Laboratories and facilities that handle hazardous substances are required to follow strict safety measures to prevent accidents or unauthorized use.

The cost of dealing with nerve agents is also high. Governments must invest in protective equipment, detection systems, medical treatments, and emergency response planning. Training personnel to handle chemical threats safely takes time and resources. In addition, the economic impact of an attack can be severe, as businesses close, transportation is disrupted, and communities struggle to recover.

Despite their harmful nature, not all chemical research is dangerous. Many scientific methods used to study chemicals have positive applications in medicine, agriculture, and industry. The problem arises when knowledge is used to create harmful substances instead of beneficial ones. This highlights the importance of ethical responsibility in science and technology.

In conclusion, nerve agents are extremely dangerous chemical weapons that affect the human nervous system in a direct and harmful way. They were developed through scientific research but were later used for military purposes, posing serious risks to humanity. While they may offer certain tactical advantages in warfare, their effects on human life, health, and the environment are severe and long-lasting. The global community has widely rejected their use, and efforts continue to eliminate them. Understanding how these weapons work and the damage they can cause helps people recognize the importance of controlling and preventing their use. The goal shared by many nations is to ensure that such weapons are never used again and that science is directed toward peaceful and helpful purposes.

4. Biological Weapons

Biological weapons are harmful agents that use living organisms or the substances they produce to cause disease and death in humans, animals, or plants. These agents include bacteria, viruses, and toxins. Unlike many other weapons, biological weapons do not rely on explosions or visible force. Instead, they spread illness, often quietly and without immediate signs. This makes them especially dangerous because people may not know they have been exposed until symptoms appear. By that time, the disease may already have spread to others, making control difficult. The effects of biological weapons can grow over time, leading to large outbreaks that affect entire communities or even multiple countries.

The idea of using disease as a weapon is not new. In ancient times, there were reports of armies trying to weaken their enemies by spreading illness. For example, contaminated materials or infected bodies were sometimes used to spread disease among opposing forces. These early methods were simple and not always effective, but they showed an early understanding that disease could be used in conflict. As science and medicine advanced, knowledge of bacteria and viruses grew. By the late nineteenth and early twentieth centuries, scientists had identified many disease-causing organisms and learned how they spread. This knowledge, while important for improving health, also opened the door to the possibility of using these agents as weapons.

During the twentieth century, several countries began to organize research into biological weapons. These programs aimed to study how diseases could be produced, stored, and spread under controlled conditions. Advances in laboratory science made it possible to grow microorganisms in larger quantities and to study their behavior in detail. Some programs focused on making agents more stable or more easily spread. Although much of this research was kept secret, over time the risks became clear—the potential for widespread harm led to international concern and efforts to limit or ban such weapons.

Biological weapons work by introducing harmful organisms or toxins into a population. Once inside the body, these agents can multiply or interfere with normal biological processes. For example, bacteria can grow and produce harmful substances, while viruses can enter cells and use them to reproduce. Toxins, which are poisonous substances produced by living organisms, can damage tissues or disrupt vital functions even in small amounts. The symptoms caused by these agents vary depending on the type of organism or toxin involved. Some diseases affect the lungs and breathing, while others may cause fever, weakness, or organ damage. In severe cases, they can lead to death.

One of the most concerning features of biological weapons is their ability to spread. Many biological agents can spread from person to person through the air, water, or direct contact. This means that a small initial release can grow into a large outbreak over time. In some cases, infected individuals may not show symptoms right away but can still pass the disease to others. This silent spread makes it hard to detect and control outbreaks in the early stages. By the time authorities recognize what is happening, the disease may have already reached many people.

From a military perspective, biological weapons have certain features that may be seen as advantages. They can affect large populations without causing immediate physical destruction to buildings or infrastructure. They can be produced in small amounts but have wide-reaching effects due to their ability to spread. In some cases, they can weaken an enemy by causing illness and reducing people's ability to work, fight, or maintain essential services. These characteristics have made biological weapons a subject of interest in some military strategies.

However, the disadvantages and dangers of biological weapons are extremely serious and far outweigh any perceived benefits. The most important issue is the loss of human life and the suffering caused by disease. Outbreaks can lead to high numbers of deaths, especially if the disease is severe or if medical care is limited. Even those who survive may suffer long-term health problems. The emotional and psychological effects are also significant, as fear and uncertainty spread along with the illness.

Another major concern is the difficulty of controlling biological outbreaks. Once a disease begins to spread, it can be hard to stop, especially if it moves quickly or if there is no effective treatment or vaccine available. Healthcare systems can become overwhelmed when large numbers of patients require care simultaneously. Hospitals may run out of space, staff, or supplies, making it harder to treat not only the disease caused by the weapon but also other medical conditions. This can increase the overall impact on society.

The effect on human civilization can be wide and long-lasting. Large outbreaks can disrupt daily life, close schools and businesses, and affect travel and trade. Economies can suffer as production slows and resources are diverted to deal with the crisis. Social systems can be strained as people try to cope with illness, loss, and uncertainty. Trust between communities and governments may weaken if people feel that information is unclear or that responses are not effective. In severe cases, entire regions may struggle to recover for many years.

Biological weapons can also affect agriculture and the environment. Some agents are designed to target animals or crops, which can lead to food shortages and economic losses. If livestock fall ill or crops fail, it can disrupt food supply chains and increase prices. This can lead to hunger and further instability. The environmental impact may include changes in ecosystems if certain species are reduced or removed due to disease.

The global nature of modern travel adds another layer of risk. Diseases can move quickly across borders as people travel from one place to another. This means that an outbreak in one area can become an international problem in a short time. Cooperation between countries is essential for monitoring, reporting, and responding to such threats. Without strong communication and coordination, the effects can grow much larger than expected.

Recognizing these dangers, the international community has taken steps to prevent the use of biological weapons. One important effort is the Biological Weapons Convention, an agreement that bans the development, production, and use of biological weapons. Many countries have joined this agreement and have committed to destroying existing stockpiles and limiting research to peaceful purposes. These efforts aim to reduce risk and promote global safety, although challenges remain in ensuring that all parties adhere to the rules.

There are also ethical concerns related to biological weapons. The use of disease as a weapon raises serious questions about human values and responsibility. Unlike many other weapons, biological agents can continue to cause harm long after they are released and can affect people who have no connection to a conflict. This makes their use widely viewed as unacceptable and inhumane.

It is important to note that the same scientific knowledge used in studying biological agents also plays a key role in improving public health. Research on bacteria and viruses has led to the development of vaccines, medicines, and improved methods for detecting and treating diseases. Laboratories around the world work to understand and control illnesses, helping to protect people rather than harm them. This shows that science itself is not the problem, but how it is used.

In conclusion, biological weapons are a serious threat because they use disease to cause harm on a large scale. Their ability to spread silently and grow over time makes them distinct from many other weapons. While they have been studied and developed in the past, their risks to human life, society, and the environment are extremely high. The global community has recognized these dangers and has taken steps to prevent their use. Understanding the nature of biological weapons helps people see why strong controls and cooperation are necessary.

The hope is that continued efforts in science, ethics, and international agreement will ensure that these weapons are never used again and that knowledge is used only to improve and protect life.

5. Dirty Bomb (Radiological Weapon)

A dirty bomb, also known as a radiological weapon, is a device that combines a conventional explosive with radioactive material. When it explodes, it does not produce a nuclear reaction, as in an atomic or hydrogen bomb. Instead, the explosion spreads radioactive particles into the air and across the surrounding area. The main purpose of such a weapon is not only to cause immediate damage from the blast, but also to contaminate the environment and create fear among people. The radioactive material can settle on buildings, roads, soil, and water, making the area unsafe for normal use for a long time. Because of this, a dirty bomb is often described as a weapon that aims to disrupt life and create long-term problems rather than cause instant, large-scale destruction.

The idea behind radiological weapons developed during the twentieth century, especially as scientists gained a better understanding of radiation and its effects on the human body. After the discovery of radioactive elements and the development of nuclear technology, it became clear that radiation could be harmful in high amounts. During the Cold War, when many countries were studying nuclear weapons, some researchers also considered the possibility of spreading radioactive material using conventional explosives. While such weapons were not widely used in warfare, the concept remained a concern because of the relative ease of obtaining certain radioactive materials from industrial, medical, or research sources. Over time, governments and international organizations have taken steps to secure these materials and reduce the risk of misuse.

A dirty bomb works simply compared to more advanced nuclear weapons. It uses a standard explosive, such as those found in conventional bombs, to break apart and scatter radioactive material. This material can come in the form of powder, pellets, or other small particles. When the bomb explodes, the blast force propels these particles into the air, where they can spread over a wide area depending on factors such as wind and weather. People in the area may breathe in the particles or come into contact with contaminated surfaces. While the explosive part of the bomb causes immediate physical damage, the radioactive part creates a longer-lasting hazard.

It is important to understand that the immediate effects of a dirty bomb are usually less severe than those of a nuclear explosion. The blast itself can cause injuries and damage similar to other types of explosives, but the radiation released is often not strong enough to cause instant death in most cases. However, the real danger lies in the long-term radiation exposure. Even low levels of radiation, if people are exposed over time, can increase the risk of serious health problems such as cancer. This makes the contamination aspect of a dirty bomb a major concern.

From a strategic point of view, one reason such weapons are considered is their ability to create widespread fear and disruption. The presence of radiation, even at low levels, can cause panic because people often associate it with severe illness and long-term danger. Large areas may need to be evacuated, and normal activities such as work, travel, and schooling may need to be suspended. Cleaning up radioactive contamination is a slow and costly process, which can keep areas closed for extended periods. This makes dirty bombs effective at causing economic damage and social disruption, even if the number of immediate casualties is limited.

However, the disadvantages of dirty bombs are significant and far-reaching. The most serious issue is the long-term health risk to people who are exposed to radiation. Depending on the level and type of radioactive material, exposure can lead to increased chances of cancer and other illnesses. People may not show symptoms right away, which can make it harder to understand the full impact at first. Over time, health problems may appear, affecting individuals and placing a burden on healthcare systems.

The environmental impact of a dirty bomb can also be severe. Radioactive particles can settle into soil and water, making land unsafe for farming and affecting food supplies. Contaminated areas may require extensive cleaning, which involves removing and safely storing affected materials. This process can take months or even years, depending on the scale of contamination. During this time, the area may remain uninhabitable, forcing people to leave their homes and communities.

The impact on human settlements is especially strong. Cities or towns affected by a dirty bomb may face long-term disruption. Buildings may need to be cleaned or even demolished if contamination is too high. Businesses may close, and property values may drop sharply. People who are forced to relocate may lose their sense of stability and connection to their community. The social and emotional effects can be deep, as individuals cope with fear, uncertainty, and loss.

Another important factor is the psychological impact. Radiation is invisible, odorless, and undetectable without special equipment. This makes it particularly frightening for many people. Even if the actual health risk is limited, the fear of exposure can lead to panic and stress. Public confidence in safety can be shaken, and rumors or misinformation can make the situation worse. Managing this fear requires clear communication and strong public health responses.

There are also economic consequences. The cost of responding to a dirty bomb attack can be very high. Emergency services, medical care, evacuation efforts, and cleanup operations all require significant resources. In addition, affected areas may lose business and tourism, leading to financial losses. Governments may need to spend large amounts of money to support recovery and rebuild affected communities.

One of the reasons dirty bombs are considered a concern is the availability of radioactive materials. While highly dangerous nuclear materials are strictly controlled, lower-level radioactive sources are used in many everyday activities, such as medical treatments, industrial equipment, and scientific research. If these materials are not properly secured, they could potentially be misused. This has led to increased efforts worldwide to track, protect, and safely manage radioactive substances.

Despite its dangers, radiation has many beneficial uses when handled properly. It is used in medicine to diagnose and treat diseases, in industry to test materials, and in research to advance scientific knowledge. The problem arises when radioactive materials are used in harmful ways. This highlights the need for careful control and responsible use of such materials.

International cooperation plays a key role in reducing the risk of radiological weapons. Many countries work together to improve security, share information, and respond to potential threats. Training programs and safety standards are used to ensure that radioactive materials are handled and stored safely. These efforts aim to prevent accidents and reduce the chance of deliberate misuse.

In conclusion, a dirty bomb is a weapon that combines a conventional explosion with the spread of radioactive material. While it may not cause immediate large-scale destruction like a nuclear bomb, its long-term effects can be serious and widespread. The contamination it creates can harm human health, damage the environment, and disrupt communities for extended periods. The fear and uncertainty it causes can be as damaging as the physical effects. Understanding how such weapons work and the risks they pose is important for improving safety and preparedness. Through strong regulations, international cooperation, and responsible handling of radioactive materials, the goal is to reduce the chances of such weapons being used and to protect people and the environment from their harmful effects.

6. Intercontinental Ballistic Missiles (ICBMs)

Intercontinental ballistic missiles, often called ICBMs, are long-range missile systems designed to travel great distances and deliver powerful warheads, including nuclear ones, from one continent to another. They are among the most advanced and dangerous delivery systems ever created. An ICBM is launched from the ground, rises high into space, and then follows a curved path back toward its target. Because of their long range, often more than 5,000 kilometers, and their very high speed, these missiles can reach distant targets in a short time, sometimes in about 30 minutes. This ability to strike faraway locations with little warning makes them a central part of modern military strategy.

The development of ICBMs has its roots in early rocket technology. During the Second World War, Germany developed the V-2 rocket, which was the first long-range guided missile used in warfare. Although it was not as advanced as modern systems, it showed that rockets could carry explosives over long distances. After the war, both the United States and the Soviet Union studied and improved this technology. As tensions between these two countries grew during the Cold War, there was a strong motivation to develop weapons capable of striking the enemy from afar. This led to rapid progress in missile design, guidance systems, and propulsion. By the late 1950s, both countries had successfully developed ICBMs, marking a major shift in global military power.

An ICBM works in several stages. It is powered by rocket engines that push it upward into the upper atmosphere and beyond. After the initial launch phase, the missile enters the midcourse phase, during which it travels through space along a calculated trajectory. During this time, the engines are no longer active, and the missile continues to move under the influence of its earlier acceleration and gravity. In the final phase, the warhead or warheads re-enter the Earth's atmosphere and descend toward the target at very high speed. Modern ICBMs often carry multiple warheads, known as MIRVs, which can be directed to different targets. This increases their effectiveness and makes them harder to defend against.

The construction and operation of ICBMs involve highly advanced technology. They require strong materials to withstand high temperatures and pressure during launch and re-entry. They also depend on precise guidance systems that use sensors, computers, and sometimes satellite navigation to ensure accuracy. These systems must work reliably under extreme conditions. Because of their complexity and destructive potential, the design and deployment of ICBMs are controlled by governments and are part of national defense programs. Detailed technical information about their construction is closely guarded.

From a military perspective, one of the main advantages of ICBMs is their deterrent role. The idea of deterrence is based on the belief that if a country can respond to an attack with equal or greater force, potential enemies may be discouraged from starting a conflict. ICBMs support this idea because they can deliver powerful warheads quickly and over long distances. Their speed and reach mean that they can strike targets even if the launching country is far away. This has been an important factor in maintaining a balance of power between nations that possess such weapons.

Another advantage is their readiness. Many ICBMs are kept in a state ready to launch on short notice. This means that they can respond quickly to threats. Some are placed in underground silos, while others may be mounted on mobile platforms or submarines. This variety makes them harder to locate and destroy, which adds to their strategic value. The ability to launch from different locations also increases flexibility in defense planning.

Despite these advantages, the disadvantages of ICBMs are very serious and far-reaching. The most important concern is the level of destruction they can cause. When equipped with nuclear warheads, ICBMs can destroy entire cities and cause massive loss of life. The blast, heat, and radiation from such an attack can have immediate and long-term effects on people and the environment. The use of multiple warheads increases the scale of destruction, enabling the targeting of several targets in a single mission.

Another major issue is the risk of accidental or unintended launch. Because ICBMs are designed for rapid response, decisions about their use may need to be made quickly. This can increase the chance of errors or misunderstandings. There have been moments in history when technical faults or false alarms created fear of an incoming attack. In such situations, the pressure to respond quickly could lead to serious consequences. This risk highlights the need for careful control, clear communication, and strong safety systems.

The impact of ICBMs on human civilization is deeply connected to the concept of nuclear warfare. Their existence has influenced international relations for decades. During the Cold War, the presence of ICBMs on both sides led to a situation often described as a balance of fear. While this may have prevented direct conflict between major powers, it also created constant tension and uncertainty. People lived with the awareness that a large-scale conflict could begin with little warning and have devastating effects.

The effect on human settlements would be severe if ICBMs were ever used in war. Cities could be destroyed in minutes, leaving behind damaged infrastructure and large numbers of casualties. Essential services such as healthcare, water supply, and transportation would be disrupted or destroyed. Survivors would face difficult conditions, including a lack of shelter, food, and medical care. In addition to the physical damage, the emotional and psychological impact on people would be profound.

Environmental damage is another serious concern. Nuclear warheads delivered by ICBMs can release radioactive materials into the air, soil, and water. This can make large areas unsafe for living and farming for many years. If many such weapons were used, the effects could extend beyond the targeted regions. Scientists have warned about the possibility of global climate change, sometimes referred to as nuclear winter, in whichatmospheric particles block sunlight. This could affect agriculture and lead to food shortages worldwide.

The financial cost of developing and maintaining ICBMs is also very high. Governments invest large amounts of money in research, production, testing, and maintenance of these systems. They also allocate resources to security and monitoring to ensure the weapons are safe and under control. These costs raise questions about how resources are used, especially in a world where many people still face basic challenges such as poverty and limited access to education and healthcare.

In response to the risks associated with ICBMs and nuclear weapons, many efforts have been made to reduce their numbers and limit their spread. International agreements have been signed to control arms and promote disarmament. These agreements aim to build trust between countries and reduce the chance of conflict. While progress has been made, many countries still maintain ICBMs as part of their defense systems, indicating that security concerns remain strong.

It is also important to note that the technology used in ICBMs has influenced other areas, such as space exploration. The same basic rocket principles are used to launch satellites and spacecraft. This shows that scientific knowledge can be used for both peaceful and military purposes, depending on how it is applied. The challenge for humanity is to guide this knowledge toward uses that benefit society rather than harm it.

In conclusion, intercontinental ballistic missiles are powerful, complex weapons capable of delivering warheads over vast distances in a short time. They were developed during a period of global tension and have played a major role in shaping modern military strategy. While they provide a sense of security through deterrence, their potential for destruction and the risks associated with their use make them a serious concern for humanity. Their impact on human life, civilization, and the environment could be severe if they were ever used on a large scale. Understanding these weapons and their effects is important for making informed decisions about their future. Many people hope that continued efforts toward cooperation, control, and peaceful use of technology will reduce the risks and help create a safer world.

7. Cruise Missiles

Cruise missiles are guided weapons designed to travel long distances and strike targets with high accuracy. Unlike ballistic missiles, which follow a high, curved path through space, cruise missiles fly at low altitude, often close to the ground or sea surface. This low flight path helps them avoid detection by radar systems and makes them difficult to track. They are powered throughout their journey by engines similar to those used in aircraft, allowing them to maintain steady flight and adjust their course as needed. Cruise missiles can carry different types of payloads, including conventional explosives or nuclear warheads, depending on their purpose.

The development of cruise missiles began during the early twentieth century, with early experiments in guided flight and uncrewed aircraft. One of the first practical examples appeared during the Second World War, when Germany developed the V-1 flying bomb. This weapon used a simple guidance system and was launched toward targets in cities such as London. Although its accuracy was limited, it showed that unmanned guided weapons could be used in warfare. After the war, advances in electronics, navigation systems, and engine technology led to the creation of more advanced cruise missiles. During the Cold War, both the United States and the Soviet Union invested heavily in these systems, improving their range, speed, and precision. Over time, cruise missiles became an important part of modern military forces.

A cruise missile works by combining propulsion, guidance, and control systems. After launch, which can take place from land, sea, or air platforms, the missile's engine keeps it moving forward at a steady speed. It uses onboard systems to guide itself toward the target. These systems may include maps stored in their memory, satellite navigation, and sensors that compare the terrain below with preloaded data. This allows the missile to follow a planned route and make adjustments during flight. Because of this ability to navigate carefully, cruise missiles can avoid obstacles and defenses while heading toward their target.

The process of creating a cruise missile involves advanced engineering and technology. It requires strong but lightweight materials, reliable engines, and highly accurate guidance systems. The design must ensure that the missile can survive different conditions, including changes in temperature, air pressure, and weather. Specialized organizations develop such weapons under strict government control. Due to their potential impact, the details of their design and production are closely guarded and not openly shared.

One of the main advantages of cruise missiles is their precision. They are designed to hit specific targets with a high degree of accuracy, thereby reducing unintended damage to surrounding areas compared to less precise weapons. This makes them useful in situations where a target must be neutralized without affecting nearby structures or populations. Their ability to fly at low altitude also makes them harder to detect and intercept. Radar systems may struggle to detect them until they are close to their targets, giving defenders less time to respond.

Another advantage is their flexibility. Cruise missiles can be launched from different platforms, including ships, submarines, aircraft, and ground-based systems. This allows military forces to use them in a variety of situations and locations. They can also carry various payloads, which increases their usefulness across different missions. Their relatively smaller size compared to some other missile systems makes them easier to transport and deploy.

However, the disadvantages of cruise missiles are significant and raise serious concerns. Despite their precision, they are still weapons designed to destroy targets, which can lead to loss of life and damage to infrastructure. If used in populated areas, even a precise strike can cause harm to civilians and disrupt daily life. The use of cruise missiles in conflict can escalate tensions and lead to further violence.

Another major concern is their stealth capability. While this is seen as an advantage from a military perspective, it also makes them dangerous because they are difficult to detect and defend against. This can increase the risk of sudden attacks, leaving little time for warning or evacuation. The presence of such weapons can create fear and uncertainty, especially in regions where conflicts are ongoing.

The impact on human civilization can be serious. The use of cruise missiles in warfare can damage critical infrastructure, including power plants, communication centers, and transportation systems. This can disrupt essential services and make it difficult for communities to function. People may lose access to electricity, clean water, and healthcare, which can worsen the effects of conflict. The destruction of infrastructure can take years to repair, slowing down recovery and development.

The environmental impact of cruise missiles should also be considered. Explosions can release harmful substances into the air and soil, affecting the health of people and animals. Damage to industrial sites or storage facilities can lead to pollution and long-term environmental problems. In some cases, fires caused by missile strikes can spread, causing further damage to natural areas.

There are also economic consequences. The development, production, and maintenance of cruise missiles require large financial investments. Governments spend significant amounts of money on these systems, which could otherwise be used for public services such as education, healthcare, and infrastructure development. In addition, areas affected by missile strikes may face economic decline as businesses close and people leave in search of safety.

Another important issue is the risk of misuse or accidental use. Because cruise missiles can be launched from different platforms and travel long distances, there is a risk that they could be used without proper authorization or in situations where mistakes occur. This could lead to unintended conflict or escalation between countries. Ensuring proper control and communication is essential to reduce these risks.

In modern times, cruise missiles have become more advanced due to technological improvements. Some are capable of flying at very low altitudes and following complex routes to avoid detection. Others are designed to travel at higher speeds, making them even harder to intercept. These developments have increased their effectiveness but also raised concerns about their impact on global security.

Despite their military role, the technology behind cruise missiles has connections to other fields. The guidance and navigation systems used in these weapons are similar to those used in civilian aircraft and navigation devices. This shows how scientific and engineering knowledge can be used both positively and negatively. The challenge is to ensure that such knowledge is applied in ways that benefit society rather than harm it.

In response to the risks posed by missile technology, countries have taken steps to curb the proliferation of such weapons. Agreements and guidelines have been developed to limit the transfer of missile technology and to promote responsible use. These efforts aim to reduce the chances of conflict and increase global stability, although challenges remain in ensuring full cooperation.

In conclusion, cruise missiles are advanced weapons that combine precision, flexibility, and stealth. They were developed over many years of technological progress and have become an important part of modern military systems. While they offer certain advantages in terms of accuracy and strategic use, their potential to cause harm and disruption makes them a serious concern. Their impact on human life, infrastructure, and the environment can be significant, especially in conflict-affected areas. Understanding how these weapons work and the risks they pose is important for making informed decisions about their use and control. Many people hope that continued efforts toward cooperation, regulation, and peaceful use of technology will help reduce the dangers associated with such weapons and contribute to a more stable and secure world.

8. Thermobaric Weapons

Thermobaric weapons, sometimes called fuel-air explosives, are a type of weapon that produces a very powerful blast by using oxygen from the surrounding air. Unlike many traditional explosives that contain both fuel and oxygen within the device, thermobaric weapons release a cloud of fuel into the air first and then ignite it. This process creates a large, intense explosion that generates intense heat and a strong pressure wave. The effect differs from a simple explosion in that it lasts longer and spreads more widely, making it especially destructive in enclosed or semi-enclosed spaces such as buildings, tunnels, and bunkers.

The history of thermobaric weapons can be traced back to the mid-twentieth century, when scientists and military engineers began studying how fuel mixtures could be used to increase the power of explosions. Early forms of this technology were explored during and after the Second World War, but more advanced versions were developed during the Cold War. Both the United States and the Soviet Union worked on improving these weapons, focusing on increasing their effectiveness against fortified positions. Over time, thermobaric weapons became part of modern military systems and have been used in several conflicts. Their development reflects ongoing efforts to create weapons capable of overcoming strong defenses and targeting hidden targets.

The basic working principle of a thermobaric weapon involves two main stages. First, the weapon releases a fine cloud of fuel particles into the air. This cloud spreads out and mixes with oxygen in the surrounding environment. Then, a second charge ignites the cloud, causing a powerful explosion. Because the fuel uses oxygen from the air, the explosion can cover a larger area than many conventional explosives. The resulting blast creates a high-pressure wave followed by a sudden drop in pressure. This combination can cause severe damage to structures and can also affect people by impacting internal organs due to the rapid change in pressure.

The construction of thermobaric weapons requires careful design and advanced technology. Engineers must ensure that the fuel is properly dispersed to form an effective cloud and that ignition occurs at the right moment. The materials used must be stable enough for storage and transport, yet reactive enough to produce the desired explosion. Specialized organizations develop these weapons under strict control, and detailed information about their design is not openly available due to their military nature.

One of the main advantages of thermobaric weapons is their ability to produce a very strong blast over a wide area. This makes them effective against targets that are difficult to destroy with standard explosives, such as bunkers, caves, and fortified buildings. The pressure wave can enter through openings such as windows and doors, reaching areas that might otherwise be protected. This allows the weapon to affect targets that are hidden or shielded. Another advantage is that they can be used in different forms, including bombs, rockets, and other delivery systems, making them adaptable to various military needs.

However, the disadvantages of thermobaric weapons are serious and raise many concerns. The most important issue is the level of destruction they cause. The intense heat and pressure can destroy buildings, collapse structures, and cause severe injuries or death to people within the affected area. Unlike some other weapons, the effects of a thermobaric explosion can extend into enclosed spaces, increasing the risk to those inside. This makes them particularly dangerous in urban environments where civilians may be present.

The impact on human health is severe. The pressure wave from the explosion can damage internal organs, even without visible external injuries. The high temperatures can cause burns and fires, increasing the danger. People who survive the initial blast may suffer from long-term physical and psychological effects. The sudden and intense nature of the explosion can create fear and trauma among affected populations.

The use of thermobaric weapons also has a strong impact on human settlements. In areas where these weapons are used, buildings can be destroyed or heavily damaged, leaving people without homes or basic services. Infrastructure such as roads, communication systems, and utilities can be disrupted, making recovery difficult. Communities may be forced to relocate, leading to long-term social and economic challenges. The destruction of homes and public spaces can take years to repair, undermining the region's stability.

Environmental effects are another concern. Explosions can release smoke, dust, and harmful substances into the air, affecting air quality. Fires caused by the heat of the explosion can spread to nearby areas, damaging vegetation and wildlife. In some cases, repeated use of such weapons can lead to long-term damage to local ecosystems. While thermobaric weapons do not produce radiation like nuclear weapons, their environmental impact can still be significant.

From a strategic perspective, thermobaric weapons are sometimes considered useful for specific military objectives. They can neutralize targets that are otherwise difficult to reach, and their powerful blast can reduce the need for repeated attacks. However, their use is often controversial because of the harm they can cause to civilians and the environment. Many people and organizations question whether the level of destruction they produce can be justified, especially in areas where non-combatants are present.

Another issue is the ethical concern surrounding their use. Weapons that cause widespread and intense damage raise questions about the limits of acceptable force in conflict. The effects of thermobaric weapons can be difficult to control, and there is a risk that they may harm people who are not directly involved in fighting. This has led to debates about their use under international humanitarian law, which aims to protect civilians during armed conflict.

The cost of developing and using thermobaric weapons is also a factor. Research, production, and deployment require resources and funding. While they may be effective in certain situations, the broader costs of rebuilding damaged areas and supporting affected populations can be very high. This highlights the long-term consequences of using such weapons beyond their immediate military purpose.

In modern times, advances in technology have made thermobaric weapons more precise and adaptable. Some systems are designed to target specific locations more accurately, which may reduce unintended damage. However, the weapon's fundamental nature remains the same, and its potential for harm remains a concern. Efforts to regulate and monitor the use of such weapons are part of ongoing discussions in international forums.

It is also important to recognize that the scientific knowledge behind thermobaric weapons is related to broader studies of energy, combustion, and materials. These fields have many positive applications, such as in industry and energy production. The challenge lies in ensuring that scientific progress is used for constructive purposes rather than destructive ones.

In conclusion, thermobaric weapons are powerful explosives that use ambient oxygen to generate intense heat and pressure. They were developed through scientific advances and have been used in military operations for specific purposes. While they offer certain advantages in terms of effectiveness against fortified targets, their destructive effects on people, buildings, and the environment are significant. Their impact on human settlements and their potential to cause suffering make them a subject of concern and debate. Understanding how these weapons work and the consequences of their use is important for informed decision-making. Many people hope that continued efforts toward responsible use, regulation, and peaceful cooperation will reduce the risks associated with such weapons and promote a safer future for all.

9. Cluster Bombs

Cluster bombs are a type of weapon designed to spread many smaller explosive units over a wide area. Instead of one large explosion, a single cluster bomb opens in the air and releases dozens or even hundreds of smaller devices, often called submunitions. These submunitions fall over a wide area and are designed to explode on impact. The goal is to affect multiple targets at once, such as groups of vehicles, troops, or equipment spread across a battlefield. Because of this wide coverage, cluster bombs are considered area weapons rather than precision weapons.

The history of cluster bombs goes back to the twentieth century, especially during the period of large-scale wars. Early forms were developed during the Second World War to increase the reach of a single bomb. Military planners wanted a method to strike several targets in a single attack without requiring many separate weapons. Over time, design improvements enabled these bombs to release more submunitions with better timing and spread. During later conflicts, including those in the second half of the twentieth century, cluster bombs were used in various regions. Their use continued into the twenty-first century, although growing concerns about their effects led to calls for restrictions and bans.

The way a cluster bomb works is based on a simple idea. The main container, often called a dispenser, is dropped or launched toward a target area. At a certain height or time, the container opens and releases its submunitions. These smaller units spread out as they fall, covering a wide area. Each submunition is designed to explode upon impact with the ground or a target. Some are made to damage vehicles, while others are intended to affect people or light structures. The design and release process requires careful timing and engineering to ensure that the submunitions spread as intended.

The production of cluster bombs involves several components, including the outer container, the submunitions, and the release mechanism. Engineers must design the system to effectively carry and distribute the smaller explosives. The materials used must be strong enough to handle transport and delivery, but also allow the container to open at the right moment. While the general concept is simple, the detailed design requires technical expertise and is carried out under strict government and military control.

One of the main advantages of cluster bombs from a military perspective is their ability to cover a large area quickly. This makes them useful in situations where targets are spread out or moving. Instead of using many separate bombs, a single cluster bomb can affect a wide zone. This can increase efficiency in certain operations and reduce the number of weapons needed for a task. Another advantage is their flexibility, as different types of submunitions can be used depending on the target.

However, the disadvantages of cluster bombs are serious and have led to strong criticism around the world. The most important problem is that not all submunitions explode as intended. A significant number may fail to detonate upon impact with the ground. These unexploded devices remain active and can act like hidden explosives. They may lie in fields, on roads, or near homes, posing a danger long after the conflict has ended. People who return to these areas may accidentally trigger them, leading to injuries or death.

The impact on civilians is one of the biggest concerns. Unexploded submunitions are often small and can be difficult to see. Children may mistake them for toys or harmless objects, which increases the risk of accidents. Farmers working in their fields, or people rebuilding their homes, may also come into contact with these devices. This creates a long-term threat that continues even after fighting has stopped. In many cases, communities are unable to use their land safely, which affects their ability to live and work normally.

The effect on human settlements can be severe. Areas affected by cluster bombs may remain dangerous for years or even decades. Roads, schools, and homes may be surrounded by unexploded devices, making daily life risky. Clearing these areas requires trained teams and specialized equipment, and the process can take a long time. During this period, people may be forced to live elsewhere or face constant danger in their own communities. This disrupts social life and can slow down recovery and development.

The environmental impact of cluster bombs is also important. Unexploded submunitions can contaminate land and prevent its use for farming or other purposes. This can affect food production and local economies. Wildlife may also be harmed, either by direct explosions or by changes in their habitat. The presence of these devices can alter the natural use of land and create long-term environmental challenges.

From a broader perspective, the use of cluster bombs raises ethical and legal questions. Because they affect wide areas and leave behind dangerous remnants, many people believe they do not meet standards meant to protect civilians during conflict. This has led to international efforts to limit or ban their use. One important step in this direction was the Convention on Cluster Munitions, an agreement that seeks to stop the use, production, and transfer of these weapons and to promote the cleanup of affected areas. Many countries have joined this agreement, although some have not, which means the issue remains complex.

Another disadvantage is the cost and effort required for cleanup. Removing unexploded submunitions is a slow and dangerous task. Teams must carefully locate and destroy each device to make the area safe. This requires training, time, and financial resources. The longer the devices remain in place, the greater the risk to people. In some regions, cleanup efforts continue for many years after the weapon's original use.

Despite these concerns, some military planners still view cluster bombs as effective in certain situations. Their ability to cover large areas quickly can be seen as useful in open battlefield conditions. However, the long-term consequences have led to increasing debate about whether their use can be justified. Many experts argue that the risks to civilians and the environment outweigh any short-term military benefits.

The impact on human civilization is significant because cluster bombs affect not only those directly involved in conflict but also future generations. The presence of unexploded submunitions can limit land use, reduce economic activity, and create ongoing fear. Communities may struggle to rebuild and return to normal life. This can lead to long-term instability and hardship.

In addition to physical harm, there is also a psychological effect. Living in an area where unexploded devices may be present can create constant stress and fear. People may avoid certain places or activities, which affects their quality of life. Children may grow up in environments where safety is uncertain, which can influence their development and well-being.

In conclusion, cluster bombs are weapons designed to disperse multiple smaller explosive submunitions over a wide area, making them effective in certain military situations. However, their tendency to leave behind unexploded submunitions poses serious, long-lasting dangers. These dangers affect civilians, communities, and the environment, often for many years after conflicts have ended. The challenges of cleanup, the risks to human life, and the impact on development have led to widespread concern and efforts to restrict their use. Understanding the nature and effects of cluster bombs helps highlight the importance of responsible decision-making and international cooperation in reducing harm and promoting safety.

10. Landmines

Landmines are explosive devices placed on or under the ground, designed to detonate when a person or vehicle comes into contact with them. They are often hidden from view, which makes them difficult to detect without special equipment. These devices are usually triggered by pressure, movement, or a tripwire. Once activated, they explode with enough force to cause serious injury or death. Landmines are used in many conflict zones to block movement, protect certain positions, or control access to land. However, their effects are not limited to wartime; they can remain active for many years after fighting has ended. The use of landmines has a long history, with early forms dating back many centuries to simple mechanical traps. However, modern landmines began to develop during the nineteenth and twentieth centuries as technology improved. During the First and Second World Wars, landmines were widely used to defend positions and slow down advancing forces. Over time, their design became more advanced, allowing them to be produced in large numbers and deployed over wide areas. During later conflicts in different parts of the world, landmines were used extensively, often without clear records of their placement. This lack of information has made them especially dangerous in the years that follow conflicts.

Landmines are generally divided into two main types. The first type is anti-personnel landmines, which are designed to injure or kill individuals. These are often triggered by the weight of a person stepping on them. The second type is anti-vehicle landmines, which are larger and require more pressure to activate. These are designed to damage or destroy vehicles, such as tanks and trucks. Both types are designed to remain hidden and ready to detonate when disturbed. The materials used in landmines can vary, but they typically include an explosive charge, a triggering mechanism, and a casing that protects the device until activation.

The process of making landmines involves combining these components in a way that ensures the device functions reliably under varying conditions. The explosive material must be stable enough for storage and transport, yet sensitive enough to detonate when triggered. The triggering mechanism must respond accurately to pressure or movement. While the basic idea is simple, the design and production of landmines require technical expertise and are conducted under controlled conditions. Details of their construction are not widely shared due to the risks involved.

From a military perspective, landmines have certain advantages. One of the main benefits is that they can remain in place for a long time without needing active control. This allows them to guard areas even when no soldiers are present. They can be used to slow down or redirect enemy forces, making it harder for them to move freely. Landmines can also protect important locations such as borders, supply routes, or defensive positions. Because they are relatively small and can be deployed in large numbers, they are considered a cost-effective means of controlling land in conflict situations.

However, the disadvantages of landmines are very serious and have led to widespread concern. The most important issue is that they do not distinguish between soldiers and civilians. Once placed, a landmine will explode when anyone steps on it, regardless of whether they are involved in the conflict. This means that innocent people, including children, farmers, and travelers, are often the victims. Injuries caused by landmines are often severe, including loss of limbs, burns, and other life-changing conditions. Many survivors require long-term medical care and support.

Another major problem is that landmines can remain active for decades. Even after a war has ended, these devices may still be hidden in the ground. Over time, their exact locations may be forgotten, especially if records were not kept. This makes it dangerous for people to return home or use their land. In many parts of the world, land that could be used for farming, building, or daily activities remains unsafe due to landmines. This creates long-term challenges for communities trying to rebuild their lives.

The impact on human settlements is significant. Villages and towns affected by landmines may be partially or completely abandoned because of safety concerns. Roads and paths may be blocked, making it difficult for people to travel or access essential services. Schools, markets, and healthcare facilities may become harder to reach. This limits opportunities for education, work, and development. In some cases, entire regions remain underused because of the risk posed by unexploded devices.

The environmental impact of landmines is also important. Land that is contaminated with mines cannot be easily used for agriculture or other purposes. This can reduce food production and affect local economies. Wildlife may also be harmed, as animals can trigger mines while moving through affected areas. The presence of landmines can change the natural use of land and disrupt ecosystems over time.

The psychological effect on people living in areas with landmines is another serious concern. The constant fear of stepping on a hidden explosive can create stress and anxiety. People may avoid certain areas, even if they are important for daily life. Children may grow up with limited freedom to play or explore their surroundings. This ongoing fear can affect mental health and reduce the overall quality of life.

Because of these serious problems, many efforts have been made to reduce the use of landmines and to remove those already in place. One important step is the international agreement known as the Ottawa Treaty, which aims to ban the use, production, and transfer of anti-personnel landmines. Many countries have joined this agreement and have committed to clearing mined areas and assisting victims. However, not all countries are part of this treaty, and landmines remain a concern in some regions.

The process of removing landmines, known as demining, is slow and dangerous. Trained teams use special equipment to detect and safely remove or destroy mines. This work requires patience and care, as even a small mistake can lead to serious injury. Demining operations can take many years, especially in areas where large numbers of mines were placed. Despite these challenges, progress has been made in clearing land and making it safe for communities to return.

The cost of dealing with landmines is also high. In addition to the cost of removal, there are costs associated with medical care for victims, rehabilitation, and support for affected communities. These costs can place a heavy burden on countries that are already recovering from conflict. At the same time, investment in clearing landmines can bring long-term benefits by allowing land to be used again for farming, housing, and development.

It is also important to note that the knowledge and technology used to make landmines can be applied more safely, for example, in construction, engineering, and resource management. The challenge is to ensure that such knowledge is used responsibly and for peaceful purposes. Education and awareness play a key role in helping people understand the risks of landmines and how to stay safe in affected areas.

In conclusion, landmines are hidden explosive devices that pose a serious threat to human life and safety. While they have been used in warfare for strategic reasons, their long-term effects on civilians, communities, and the environment are severe. They can remain active for many years, causing harm long after conflicts have ended. The presence of landmines can prevent development, disrupt daily life, and create lasting fear. Efforts to ban their use and remove existing mines are important steps toward reducing their impact. Understanding the dangers of landmines helps highlight the need for continued action to protect people and restore affected areas.

11. Napalm

Napalm is a type of incendiary weapon that is designed to burn very intensely and spread fire over a wide area. It is not a single chemical substance, but a mixture that turns fuel into a thick, sticky gel. This gel can cling to surfaces such as buildings, vegetation, and even human skin. When ignited, it burns at very high temperatures and can continue to burn for a long time. Because of its sticky nature, it is very difficult to remove once it comes into contact with a surface. This makes napalm especially dangerous, as it can cause deep burns and start large fires that are hard to control.

The development of napalm began during the Second World War. Scientists in the United States worked on ways to improve the effectiveness of flammable weapons. They wanted a substance that would burn longer and spread more easily than regular liquid fuel. By combining fuel with certain thickening agents, they created a gel that could stick to targets and burn more effectively. The name "napalm" comes from the chemicals used in early versions of the mixture. After its development, napalm was used in several military operations during the war. It was later used in other conflicts, including the Korean War and the Vietnam War, where its effects became widely known worldwide.

Napalm works by combining fuel with a thickening agent that changes its consistency from liquid to gel. This gel is then placed in bombs or other delivery systems. When the weapon is released and ignited, the gel spreads over the target area and burns intensely. The burning fuel consumes oxygen and produces high heat, which can ignite nearby materials and cause fires to spread. The gel's thickness allows it to stay in place and continue burning, rather than flowing away like a liquid. This increases the duration and impact of the fire.

The production of napalm involves chemical processing and careful mixing of materials to achieve the desired properties. The mixture must be stable enough for storage and transport, but also capable of igniting easily when needed. The design of the delivery system must ensure that the gel is effectively spread over the target area. Because of the risks involved, the development and handling of such materials are controlled by governments and military organizations. Detailed methods of production are not openly shared due to safety and security concerns.

From a military perspective, napalm has certain advantages. One of the main benefits is its ability to create large fires capable of destroying structures, vegetation, and supplies. It can be used to clear areas, disrupt enemy positions, or deny access to certain locations. The intense heat can damage equipment and make areas difficult to occupy. Its ability to stick to surfaces increases its effectiveness compared to liquid fuels, as it remains in place and continues to burn.

However, the disadvantages of napalm are severe and have led to strong criticism. The most serious issue is the harm it causes to people. When napalm comes into contact with the human body, it can cause deep and painful burns. Because it sticks to the skin, it is very difficult to remove, and the burning can continue for a long time. These injuries are often life-threatening and can lead to long-term physical and emotional suffering for survivors. The use of napalm in populated areas can result in high numbers of civilian casualties.

Another major concern is the uncontrollable nature of fire. Once napalm ignites, the flames can spread beyond the intended target. This can lead to large fires that destroy homes, forests, and other important areas. In some cases, entire neighborhoods or villages can be affected. The spread of fire can also make it difficult for emergency services to respond, increasing the damage and risk to people.

The impact on human settlements is significant. Areas affected by napalm may suffer from widespread destruction of buildings and infrastructure. Homes, schools, and hospitals can be damaged or destroyed, leaving people without shelter or essential services. The recovery process can take a long time, as rebuilding requires resources and effort. Communities may be forced to relocate, disrupting social networks and economic activities.

The environmental impact of napalm is also important. Fires caused by napalm can destroy vegetation and wildlife, leading to loss of habitats and changes in ecosystems. The burning of large areas can release smoke and pollutants into the air, affecting air quality and health. Soil can be damaged by intense heat, making it less suitable for farming. In some cases, the effects on the environment can last for many years.

The psychological impact of napalm is another serious concern. The sight and experience of intense fires can cause fear and trauma among those affected. People who survive such events may carry long-term emotional scars. The use of napalm has also influenced public opinion about warfare, as images and reports of its effects have raised awareness about the human cost of such weapons.

From an ethical point of view, the use of napalm has been widely debated. Many people argue that weapons that cause severe suffering and cannot be easily controlled should not be used, especially in areas where civilians are present. This has led to discussions about the rules of war and the need to protect non-combatants. International agreements have been developed to limit the use of certain types of incendiary weapons, including napalm, particularly in civilian areas.

The cost of using napalm is not only measured in financial terms but also in human and environmental damage. While the materials used to create napalm may not be as expensive as some advanced weapons, the long-term effects of its use can be very costly. Rebuilding damaged areas, providing medical care for victims, and restoring the environment all require significant resources. These costs can place a heavy burden on affected communities and governments.

It is also important to note that the scientific knowledge used to develop napalm has applications beyond its development. The study of fuels, combustion, and materials is important for industries such as energy, transportation, and manufacturing. When used responsibly, this knowledge can lead to positive developments. The challenge is to ensure that it is not used in ways that cause harm.

In conclusion, napalm is a powerful incendiary weapon that uses a sticky, flammable gel to create intense and long-lasting fires. It was developed during a time of war and has been used in several conflicts, where its effects have been both effective and devastating. While it offers certain military advantages, its impact on human life, communities, and the environment is severe. The injuries it causes, the destruction it creates, and the fear it spreads make it one of the most controversial weapons in modern history. Understanding how napalm works and the consequences of its use is important for making informed decisions about its role in warfare. Many people hope that continued efforts to limit harmful weapons and promote peaceful solutions will reduce the risks associated with such technologies and help build a safer future.

12. White Phosphorus

White phosphorus is a chemical substance that has been used in military settings for many years because of its ability to ignite quickly and produce intense heat and thick smoke. It is a form of the element phosphorus that reacts strongly with oxygen in the air. When exposed to air, it can ignite spontaneously without a spark or flame. This makes it useful in certain situations where rapid ignition is needed. It burns at very high temperatures, producing a bright flame and dense white smoke. Because of these properties, white phosphorus has been used in various ways, including creating smoke screens to conceal movement and as an incendiary to start fires.

The history of white phosphorus in warfare goes back to the early twentieth century, when advances in chemistry made it easier to produce and handle such materials. During the First World War, armies began to use chemical substances more widely, including those that could create smoke or fire. White phosphorus became one of the materials used for these purposes. Its ability to produce thick smoke made it useful for hiding troop movements and marking targets. Throughout the Second World War and subsequent conflicts, its application persisted and grew. It was used in various forms, including shells, grenades, and bombs. Over time, its effects became better understood, and its use became a subject of debate due to the harm it can cause.

White phosphorus reacts chemically with oxygen. When it comes into contact with air, it oxidizes rapidly, producing heat and light. This reaction continues as long as oxygen is present, meaning the substance will keep burning until it is either completely consumed or deprived of oxygen. The smoke produced during this process consists of fine particles that can spread over an area, reducing visibility. This makes it useful for creating smoke screens in military operations. However, the same reaction that makes it useful also makes it dangerous, especially when it comes into contact with living tissue.

The preparation and use of white phosphorus involve chemical processes that are carefully controlled. It is usually stored in a way that prevents exposure to air, often under water or in sealed containers. When used in weapons, it is placed in a delivery system that releases it into the air or onto a target area. Upon release, it comes into contact with oxygen and ignites. The design of these systems requires technical knowledge and careful handling to ensure safety during storage and transport. Because of the risks involved, the details of production and handling are regulated and not openly shared.

From a military perspective, white phosphorus has certain advantages. One of its main uses is in creating smoke screens. The thick smoke can block visibility, making it harder for opponents to see movements or positions. This can provide cover for troops or vehicles and can be used to signal or mark locations. Another advantage is its ability to start fires quickly. When used as an incendiary substance, it can ignite materials such as wood, fabric, or fuel, which can be useful in certain operations. Its rapid ignition and high temperature make it effective in these roles.

However, the disadvantages and risks of white phosphorus are significant. The most serious concern is the harm it can cause to people. When it comes into contact with the skin, it can cause deep, severe burns. Because it continues to burn as long as it is exposed to air, it can be very difficult to extinguish. Even if the flames appear extinguished, the substance can reignite if exposed to oxygen again. This makes injuries caused by white phosphorus particularly dangerous and difficult to treat. Medical care for such burns is complex and may require long-term treatment.

Another major issue is the difficulty of controlling its effects. Once white phosphorus is released and ignited, it can spread over an area and continue burning. This can lead to fires that are hard to manage, especially in places with flammable materials. In urban areas, this can damage buildings and infrastructure. The spread of fire can also pose a risk to people who are not directly involved in the conflict, including civilians.

The impact on human settlements can be severe. The use of white phosphorus in populated areas can destroy homes, public buildings, and essential services. Fires can damage or destroy property, leaving people without shelter. The smoke produced can reduce visibility and create confusion, making it harder for people to escape or for emergency services to respond. The long-term effects may include the need to rebuild damaged areas and support affected communities.

The environmental impact is also important to consider. Fires caused by white phosphorus can damage vegetation and wildlife. The smoke released into the air can affect air quality and may harm people who inhale it. In some cases, residues from the substance can remain in the environment for a period of time, although they usually break down eventually. The overall effect depends on the scale of use and the conditions in the affected area.

The psychological impact of white phosphorus is another factor. The intense flames and thick smoke can create fear and panic among people in the affected area. The knowledge that the substance can continue burning and reignite adds to this fear. Survivors of such incidents may experience long-term emotional effects, including stress and anxiety. These effects can influence communities and their sense of safety.

The use of white phosphorus has also raised ethical and legal questions. While it has legitimate uses, such as creating smoke screens, its use as an incendiary substance in areas where civilians are present is highly controversial. International laws and agreements aim to limit the use of weapons that can cause unnecessary suffering or that cannot be controlled effectively. The debate over white phosphorus reflects broader concerns about how to balance military needs with the protection of human life.

The cost of using white phosphorus is not limited to the materials themselves. There are also costs related to medical care for those injured, rebuilding damaged areas, and addressing environmental effects. These costs can be high and may continue long after the initial use of the substance. This highlights the importance of considering the long-term consequences of its use.

It is also worth noting that phosphorus has many beneficial uses outside of warfare. It is an essential element in agriculture, where it is used in fertilizers to support plant growth. It is also used in industry and manufacturing. This shows that the same chemical can have both helpful and harmful uses, depending on how it is applied. The challenge is to ensure that such materials are used responsibly and safely.

In conclusion, white phosphorus is a chemical substance that ignites on contact with air and produces intense heat and thick smoke. It has been used in military operations for creating smoke screens and as an incendiary material. While it offers certain advantages in terms of rapid ignition and effectiveness, its risks are serious. The burns it causes, the fires it can start, and the difficulty in controlling it make it a dangerous substance. Its impact on people, communities, and the environment can be severe, especially when used in populated areas. Understanding its properties and effects helps to highlight the importance of careful regulation and responsible use. Many people believe that efforts should continue to limit its harmful applications and ensure that safety and human well-being are prioritized.

13. Cyber Weapons

Cyber weapons are tools created in the digital world to attack computer systems, networks, and data. Unlike traditional weapons, they do not rely on physical force or visible destruction. Instead, they use software, code, and network access to disrupt or damage important systems. These weapons can target many types of infrastructure, including power grids, communication networks, banking systems, transportation systems, and healthcare services. Because modern societies depend heavily on digital technology, cyber weapons have become a serious concern in recent years. Their effects can spread quickly and affect large numbers of people without any physical explosion or visible sign of attack.

The development of cyber weapons is closely linked to the growth of computers and the internet. In the early days of computing, systems were mostly isolated, and the idea of attacking them was limited. As networks expanded and the internet connected the world, new possibilities emerged. By the late twentieth century, simple forms of malicious software began to appear, often created by individuals experimenting with code. Over time, these tools became more advanced and were used for harmful purposes such as stealing data or disrupting services. In the twenty-first century, governments and organized groups began developing more sophisticated cyber tools for strategic purposes. These tools could target critical systems and cause large-scale disruption, marking the rise of cyber warfare as a new form of conflict.

Cyber weapons work by exploiting weaknesses in computer systems. These weaknesses, often called vulnerabilities, can exist in software, hardware, or network configurations. Once a vulnerability is found, a cyber weapon can use it to gain access to a system, change its behavior, or shut it down. Some cyber tools are designed to spread from one system to another, increasing their impact. Others focus on specific targets, such as a particular organization or type of infrastructure. The effects can range from slowing down systems to completely disabling them. In some cases, cyber attacks can also manipulate data, leading to confusion and incorrect decisions.

The creation of cyber weapons involves advanced knowledge of programming, networks, and system design. Developers study how systems work and identify points where they can be accessed or disrupted. They then design software that can use these points to achieve a specific goal. This process requires skill and resources, especially for large-scale or targeted attacks. Because of the potential risks, information about such tools is often kept secret, and governments or specialized groups control their development. Unlike physical weapons, cyber weapons can be created and modified quickly, making them adaptable yet harder to regulate.

One of the main advantages of cyber weapons is that they can operate without physical presence. An attacker can launch an operation from a distant location, often across borders, without entering the target area. This reduces the risk of confrontation and makes it harder to identify the source of the attack. Another advantage is their ability to target specific systems with precision. For example, a cyberattack can target a single network or facility without affecting surrounding areas. This level of control can be seen as useful in certain strategic situations.

Cyber weapons can also be less costly to develop compared to traditional military equipment. They do not require large factories or physical materials in the same way as conventional weapons. This means that even smaller groups may be able to create or use them. In addition, cyber operations can be carried out quickly, sometimes within minutes, and can be repeated or adjusted as needed. These features make cyber weapons a flexible and powerful tool in modern conflict.

However, the disadvantages and risks of cyber weapons are serious and wide-reaching. One major concern is the impact on essential services.

Modern cities rely on digital systems to manage electricity, water supply, transportation, and communication. A cyber attack on these systems can cause widespread disruption. For example, if a power grid is affected, entire regions may lose electricity, leading to problems in homes, hospitals, and businesses. Such disruptions can create confusion and difficulty in daily life.

Another issue is the indirect harm caused by cyber attacks. While they may not cause physical destruction in the traditional sense, their effects can still lead to real-world consequences. For instance, if healthcare systems are disrupted, patients may not receive timely care. If financial systems are targeted, people may lose access to their money or face economic instability. These effects can spread across society and affect large numbers of people.

The difficulty of identifying the source of a cyber attack is another challenge. Attackers can hide their identities using various techniques, making it hard to determine who is responsible. This can lead to misunderstandings and increase tension between countries or groups. Without clear information, responding to an attack becomes more complex and may lead to further conflict.

The impact on human civilization is significant because cyber weapons affect the systems that support modern life. As societies become more connected through technology, the potential damage from cyber attacks increases. A large-scale attack could disrupt communication, trade, and public services, affecting the stability of entire regions. This creates a new type of risk that did not exist in the past, in which damage is not always visible but is still deeply felt.

Human settlements are also affected by cyber weapons. Cities depend on digital systems for traffic control, public safety, and resource management. If these systems fail, it can lead to confusion and safety risks. For example, traffic signals may stop working, causing accidents or delays. Emergency services may face difficulties in responding to incidents. Over time, repeated disruptions can reduce confidence in public systems and make daily life more challenging.

The economic impact of cyber weapons is another important factor. Businesses rely on digital systems for operations, communication, and data storage. A cyberattack can disrupt these activities, leading to financial losses. In some cases, sensitive information may be stolen, harming individuals and organizations. The cost of recovering from an attack, including repairing systems and improving security, can be very high.

Cyber weapons also raise ethical and legal questions. Because they can affect civilians and essential services, their use in conflict is a subject of debate. International law is still developing in this area, as countries seek to understand how to apply existing rules to digital operations. There is a need for clear guidelines to ensure that actions in cyberspace do not cause unnecessary harm.

Another point of concern is the quick evolution of technology. As systems become more advanced, new vulnerabilities may appear. This creates a constant need for security updates and monitoring. At the same time, attackers continue to develop new methods, leading to an ongoing cycle of action and response. This makes it difficult to achieve complete safety.

Despite these challenges, efforts are being made to improve cybersecurity and reduce risks. Governments, organizations, and experts work together to strengthen systems, share information, and respond to threats. Education and awareness are also important, as individuals play a role in protecting their own data and devices. By improving security practices, it is possible to reduce the impact of cyber attacks.

It is important to understand that the technology used in cyber weapons also has many positive uses. The same skills and tools can be used to protect systems, improve communication, and support innovation. The essential difference resides in the application of the technology. Responsible use of digital knowledge can lead to progress and benefit society.

In conclusion, cyber weapons are a modern form of conflict that uses digital tools to disrupt and damage computer systems and infrastructure. They have developed alongside technological advances and have become a significant part of global security concerns. While they offer certain advantages, such as precision and remote operation, their potential to disrupt essential services and affect large populations makes them highly dangerous. Their impact on human life, economic stability, and social systems can be serious, even without physical destruction. Understanding these weapons and their effects is important for building stronger defenses and promoting responsible use of technology. Many people hope that cooperation and awareness will help reduce the risks and ensure that digital tools are used for the benefit of humanity rather than harm.

14. Autonomous Weapon Systems (AI Weapons)

Autonomous weapon systems, often called AI weapons, are machines that can carry out military tasks without direct human control at the moment of action. These systems use artificial intelligence to observe their surroundings, process information, and decide what to do next. In some cases, they can identify targets and carry out attacks based on pre-set rules and learned patterns. Unlike traditional weapons, which rely on human operators to guide them at every step, autonomous systems are designed to operate independently once activated. This ability to operate independently makes them one of the most debated developments in modern technology.

The idea of machines performing tasks without human control is not entirely new. Early examples can be seen in automated systems used in industry and transportation. However, the use of such systems in warfare began to take shape in the late twentieth and early twenty-first centuries, as computers became more powerful and better able to process large amounts of data. Advances in sensors, robotics, and machine learning made it possible to create systems that could respond to changing conditions in real time. Military research programs began to explore how these technologies could be used for defense and security purposes. Over time, this led to the development of systems capable of performing surveillance, navigation, and even targeting with reduced human involvement.

Autonomous weapons work by combining several key components. They use sensors, such as cameras, radar, or other detection tools, to gather information about their environment. This information is then processed by computer systems that use algorithms to interpret what is happening. Based on this analysis, the system can make decisions about movement, target selection, and action. Some systems are designed to operate under human supervision, where a person can intervene if needed. In contrast, others are built to operate with very little or no direct control once deployed. The level of autonomy can vary, but the defining feature is the ability to act without continuous human input.

The creation of autonomous weapon systems involves advanced research in artificial intelligence, robotics, and engineering. Developers design software that can recognize patterns, learn from data, and make decisions. They also build physical platforms, such as drones, ground vehicles, or other machines, that can carry out these decisions. The process requires careful testing to ensure that the system behaves as intended under different conditions. Because of the complexity and potential risks, the development of such systems is usually carried out by governments and specialized organizations. Detailed technical information is often kept confidential.

One of the main advantages of autonomous weapons is their ability to operate quickly and efficiently. Machines can process information faster than humans and respond to environmental changes without delay. This can be useful in situations where rapid decisions are needed. Another advantage is that they can reduce the risk to human soldiers, as they can be sent into dangerous areas instead of people. Autonomous systems can also operate for long periods without fatigue, which may improve performance in certain tasks such as surveillance or patrol.

In addition, these systems can be designed to carry out precise actions based on data and rules. This has the potential to reduce unintended damage if the system is accurate and well-designed. For example, a system could be programmed to avoid certain areas or to act only under specific conditions. This level of control is seen by some as a benefit, especially in complex environments where human decision-making may be slower or affected by stress.

However, the disadvantages and risks of autonomous weapons are significant and have led to widespread concern. One of the most important issues is the lack of human judgment at the moment of action. Human decision-making includes ethical reasoning, understanding of context, and the ability to adapt to unexpected situations. Machines, on the other hand, rely on data and algorithms, which may not fully capture the complexity of real-world scenarios. This raises the risk that a system could make incorrect decisions, leading to unintended harm.

Another major concern is accountability. If an autonomous system makes a mistake, it can be difficult to determine who is responsible. Questions arise about whether the responsibility lies with the developer, the operator, or the organization that deployed the system. This lack of clear responsibility creates challenges for legal and ethical frameworks. It also makes it harder to ensure that such systems are used in ways that respect human rights and international law.

The impact on humanity can be profound. The use of autonomous weapons may change the nature of conflict by reducing the need for direct human involvement. This could make conflicts more likely to begin, as the immediate risk to human soldiers is lower. At the same time, the potential for harm remains high, especially if systems are used without careful control. The presence of such weapons may also heighten fear and uncertainty, as people may feel that machines rather than humans are making life-and-death decisions.

Human civilization may also be affected in broader ways. The development and use of autonomous weapons could lead to new forms of competition between countries, as each seeks to advance its technology. This could increase tension and reduce trust. There is also concern about the spread of such technology, as it may become available to groups that do not follow international rules. This could lead to misuse and increase the risk of harm.

The effect on human settlements is another important consideration. Autonomous systems could be used in urban areas, where the presence of civilians complicates decision-making. If a system fails to identify a target correctly or misinterprets a situation, it could cause damage to homes, infrastructure, and public spaces. This could disrupt daily life and create long-term challenges for affected communities.

There are also technical risks associated with autonomous weapons. Systems that rely on software and networks can be vulnerable to errors, malfunctions, or external interference. For example, a system could be affected by incorrect data, leading to wrong decisions. It could also be targeted by cyberattacks, which might alter its behavior or turn it off. These risks highlight the importance of strong security and testing, but they cannot be eliminated.

The ethical concerns surrounding autonomous weapons are among the most widely discussed issues. Many experts believe that decisions about the use of force should always involve human judgment. They argue that removing humans from this process raises serious moral questions. There is an ongoing debate about whether such systems should be limited or banned. Some organizations and countries have called for international agreements to regulate their development and use.

Despite these concerns, research in artificial intelligence continues to grow, and the same technologies used in autonomous weapons have many positive applications. AI can be used in healthcare, education, transportation, and many other fields to improve efficiency and quality of life. This shows that the technology itself is not harmful, but its use in certain contexts can create risks.

Efforts are being made to address the challenges posed by autonomous weapons. Discussions at international levels aim to create guidelines and rules for their use. These efforts focus on ensuring that human oversight remains part of decision-making and that systems are designed with safety and responsibility in mind. While progress has been made, much work remains to reach a global agreement.

In conclusion, autonomous weapon systems represent a significant change in the way technology is used in conflict. They combine artificial intelligence and robotics to create systems that can act without direct human control. While they offer advantages such as speed, efficiency, and reduced risk to human operators, they also raise serious concerns about safety, accountability, and ethics. Their impact on humanity, society, and global stability could be far-reaching. Understanding these systems and their effects is important for guiding their development and use. Many people believe that careful regulation, strong oversight, and international cooperation are necessary to ensure that such technologies do not cause harm and are used responsibly.

15. Drone Swarms

Drone swarms are groups of uncrewed aerial vehicles, commonly called drones, that work together in a coordinated way to carry out tasks. Unlike a single drone, a swarm operates as a connected system where each unit communicates with others and follows shared instructions. These drones can move together, spread out, regroup, and adjust their actions in response to environmental changes. In military use, drone swarms can be used for surveillance, monitoring, and in some cases, carrying out attacks. Their ability to act as a group makes them more effective and harder to stop than individual drones.

The development of drone swarms is closely linked to advances in robotics, communication systems, and artificial intelligence. Early drones were controlled directly by human operators and could perform only limited tasks. As technology improved, engineers began to explore how multiple drones could work together. Research in fields such as swarm intelligence, which studies how groups of animals, such as birds and insects, move in coordination, has helped guide this development. By the early twenty-first century, improvements in sensors, wireless communication, and computing power made it possible to create drone systems that could share information and act together in real time. Military and research organizations around the world have since invested in this technology.

Drone swarms operate through a combination of communication and decision-making systems. Each drone in the swarm is equipped with sensors to observe its surroundings and software to process information. The drones share data, often through wireless networks, and use this information to coordinate their actions. A central system guides some swarms, while others operate in a more distributed manner, in which each drone follows simple rules that lead to complex group behavior. This allows the swarm to adapt to changes, such as obstacles or threats, without needing constant human control.

The creation of drone swarms involves designing both the individual drones and the system that connects them. Each drone must be able to fly, sense its environment, and communicate with others. The software that controls the swarm must be carefully designed to ensure that the drones work together effectively and safely. This requires testing under many conditions to ensure the system behaves as expected. Because of the potential risks, the development of such systems is usually carried out under strict control, and detailed technical information is not widely shared.

One of the main advantages of drone swarms is their ability to overwhelm defenses. A single drone can be detected and stopped more easily, but a large number of drones approaching at once can be much harder to defend against. Even if some drones are intercepted, others may still reach their target. This makes swarms a powerful tool in situations where defenses are strong. Another advantage is their flexibility. Swarms can be used for different tasks, such as gathering information, mapping areas, or delivering payloads. They can also cover large areas quickly by spreading out and then regrouping when needed. Drone swarms can also reduce risk to human operators. Because they are unmanned, they can be sent into dangerous areas without putting people directly in harm's way. This can be useful for surveillance in conflict zones or for assessing situations where it would be unsafe for humans to go. In addition, drones can operate continuously for extended periods, depending on their design, which allows them to monitor areas over time.

However, the disadvantages and risks of drone swarms are serious and raise important concerns. One major issue is the potential for loss of control. Because swarms rely on communication and software, problems such as signal interference, technical errors, or system failures could affect their behavior. If coordination is lost, drones could act unpredictably, leading to unintended consequences. This risk increases as the size and complexity of the swarm grow.

Another concern is the difficulty of defense. Traditional defense systems are often designed to deal with single or limited threats. A large swarm can approach from different directions and at different speeds, making it harder to respond effectively. This can increase the likelihood of damage to targets and reduce defenders' ability to protect important areas.

The impact on human life can be significant, especially if drone swarms are used in populated areas. Their ability to move quickly and act in coordination means they can carry out actions in a short time, leaving little opportunity for people to react. If used in attacks, they could harm civilians and damage infrastructure. Even when used for surveillance, their presence can create a sense of fear and lack of privacy.

Human settlements may be affected in several ways. In cities, drone swarms could disrupt transportation, communication, and public services. For example, they could interfere with the airspace used by other aircraft or affect signals in communication systems. In the event of an attack, buildings and essential facilities could be damaged, disrupting daily life. The need to protect against such threats may also lead to increased security measures, which can change how people live and move within their communities.

The economic impact of drone swarms is another factor to consider. Developing and maintaining these systems requires investment in technology, research, and training. At the same time, defending against them also requires resources, as new systems must be developed to detect and stop swarms. If used in conflict, drone swarms can cause damage that leads to financial losses, including repair and recovery costs.

There are also ethical concerns related to drone swarms. The use of automated systems that can act with limited human control raises questions about responsibility and decision-making. If a swarm causes harm, it may be difficult to determine who is accountable. There is also concern about the use of such systems in ways that may not fully consider civilian safety. These issues have led to discussions about the need for clear rules and guidelines.

Another important point is the possibility of misuse. Because drone technology is becoming more accessible, there is a risk that groups or individuals could use such systems outside of formal military control. This increases the importance of regulation and monitoring to prevent harmful use. Ensuring that these technologies are used responsibly is a challenge that requires cooperation between countries and organizations.

Despite these risks, the technology behind drone swarms has positive applications as well. Similar systems can be used for search-and-rescue operations, environmental monitoring, and disaster response. For example, swarms of drones can be used to locate people in difficult terrain or to assess damage after natural disasters. This shows that the same technology can be used for beneficial purposes when applied responsibly.

Efforts are being made to address the challenges posed by drone swarms. Researchers are working on systems to detect and counter swarms, as well as ways to improve safety and control. Discussions at international levels aim to create guidelines for the use of such technologies. These efforts focus on balancing the benefits of innovation with the need to protect people and maintain stability.

In conclusion, drone swarms are a modern development that combines robotics, communication, and artificial intelligence to create coordinated groups of drones. They offer advantages such as flexibility, speed, and the ability to operate without direct human presence. However, their potential to overwhelm defenses, cause harm, and create new security challenges makes them a serious concern. Their impact on human life, infrastructure, and global stability depends on how they are used and controlled. Understanding drone swarms and their effects is important for guiding their development and ensuring they are used in ways that prioritize safety and responsibility.

16. Electromagnetic Pulse (EMP)

Electromagnetic pulse weapons, often called EMP weapons, are devices designed to release a burst of energy that can damage or turn off electronic systems. This burst is not visible like fire or an explosion, but it travels as a powerful wave of electromagnetic energy. When this energy reaches electronic equipment, it can overload circuits and cause devices to stop working. Because modern societies depend heavily on electronics, the effects of such a weapon can be far-reaching and severe. An EMP does not directly harm people in the same way as traditional weapons. Still, it can affect the systems that support daily life, such as electricity, communication, transportation, and healthcare.

The idea behind electromagnetic pulses comes from the study of electricity and magnetism. Scientists have known for a long time that strong electrical currents can create magnetic fields, and that sudden changes in these fields can produce powerful energy waves. During the mid-twentieth century, researchers began to notice that certain high-energy events, including nuclear tests, could create strong electromagnetic effects. These effects could interfere with electronic systems even at a distance. During the Cold War, both the United States and the Soviet Union studied these effects more closely. This led to the development of weapons that could use electromagnetic pulses intentionally to disrupt enemy systems.

There are two main types of EMP weapons. One type is linked to nuclear explosions. When a nuclear device is detonated at high altitude, it can produce a large electromagnetic pulse that spreads over a wide area. This type of EMP can affect entire regions, depending on the explosion's altitude and other factors. The second type is non-nuclear EMP, which uses specialized equipment to generate electromagnetic energy without a nuclear reaction. These devices are smaller and more focused, but they can still cause significant damage to electronic systems within their range.

EMP weapons work by emitting a sudden, intense burst of energy into the environment. This energy can travel through the air and also through wires, cables, and other conductive paths. When it reaches electronic devices, it can create high voltages inside circuits. These voltages can damage components, erase data, or cause systems to shut down completely. Some systems may recover after the event, but others may suffer permanent damage and require replacement.

The creation of EMP weapons involves advanced knowledge of physics and engineering. Scientists and engineers design systems that can generate and release electromagnetic energy in a controlled way. For nuclear-based EMP, the effect results from the nuclear explosion itself and interactions in the upper atmosphere. For non-nuclear EMP, devices may use high-power electrical systems to produce the pulse. Because of the risks and complexity, the development of such weapons is usually carried out under strict control, and detailed information is not widely shared.

From a military perspective, EMP weapons have certain advantages. One of the main benefits is their ability to turn off systems without causing direct physical destruction. For example, an EMP attack could shut down power grids, communication networks, or control systems without damaging buildings. This can reduce visible damage while still achieving a strategic objective. Another advantage is the wide area that can be affected, especially in the case of high-altitude nuclear EMP. A single event can impact a large region, making it a powerful tool for disruption.

EMP weapons can also be used to target specific systems. Non-nuclear EMP devices can be designed to focus on particular locations or types of equipment. This allows for more precise use in certain situations. In addition, because EMP effects propagate quickly, the impact can be almost immediate, leaving little time to respond.

However, the disadvantages and risks of EMP weapons are very serious. One major concern is the impact on essential infrastructure. Modern life depends on electricity and electronic systems. Power grids control the supply of electricity to homes, hospitals, and businesses. Communication systems connect people and allow coordination of services. Transportation systems rely on electronic controls for safety and efficiency. If these systems are disrupted, the effects can spread quickly and affect large populations.

The impact on human life can be indirect but still severe. For example, if hospitals lose power, medical equipment may stop working, putting patients at risk. If water supply systems fail, access to clean water may be affected. If communication systems are disrupted, emergency services may find it difficult to respond to incidents. These effects can create serious challenges, especially if the disruption is prolonged.

Human settlements can be deeply affected by EMP events. Cities that rely heavily on technology may experience widespread blackouts and service outages. Traffic systems may stop working, leading to confusion and accidents. Businesses may close due to a lack of power or communication, resulting in economic losses. Daily life can become difficult as people struggle to access basic needs. In rural areas, the impact may also be significant, especially if communication and supply systems are affected.

The economic consequences of EMP weapons can be very large. Repairing damaged systems and restoring services can take time and resources. In some cases, electronic components may need to be replaced, which can be costly and time-consuming to obtain. The disruption of business activities can lead to financial losses for individuals and organizations. Over time, these effects can slow down economic growth and recovery.

Another important issue is the difficulty of protection. While some systems can be shielded against electromagnetic effects, it is not always possible to protect everything. Building strong defenses requires planning and investment. Even with protective measures, some systems may still be affected. This makes it challenging to fully prepare for such events.

There are also concerns about escalation. The use of EMP weapons, especially those linked to nuclear devices, can increase tension between countries. Because the effects can cover large areas, they may affect not only the intended target but also neighboring regions. This can lead to misunderstandings and increase the risk of further conflict.

The environmental impact of EMP weapons is generally less direct than that of explosive weapons, but it can still be significant. For example, long-term power outages can affect water treatment, waste management, and other systems that protect the environment. This can lead to secondary effects such as pollution or resource shortages.

Despite these risks, the science behind electromagnetic pulses has positive applications. Understanding electromagnetic effects helps design better electronic systems and improve safety. For example, engineers use this knowledge to protect devices from natural events such as lightning or solar storms, which can also produce electromagnetic disturbances. This shows that the same knowledge can be used for both protective and harmful purposes.

Efforts are underway to improve resilience to EMP events. Governments and organizations are working to strengthen infrastructure, develop backup systems, and improve emergency response plans. These measures aim to reduce the impact of disruptions and ensure that essential services can continue even in difficult conditions. International discussions also focus on reducing the risks associated with such weapons and promoting stability.

In conclusion, electromagnetic pulse weapons are technologies that can disrupt or turn off electronic systems by releasing a powerful burst of energy. They have developed alongside advances in science and have become an important topic in modern security discussions. While they offer certain advantages in terms of wide-area impact and reduced physical destruction, their potential to disrupt essential services makes them highly dangerous. Their effects on human life, infrastructure, and the economy can be severe, even without visible damage. Understanding how these weapons work and the risks they pose is important for building stronger systems and ensuring safety. Many people believe that careful management, cooperation, and responsible use of technology are necessary to reduce the dangers associated with EMP weapons and to protect society from their effects.

17. Sniper Rifles

Sniper rifles are firearms designed for very high accuracy over long distances. They are built to allow a trained user to hit a specific target with precision, often from far away. Military forces and law enforcement units use these rifles in situations requiring careful aiming and controlled use of force. A sniper rifle differs from a standard rifle in its strong focus on stability, accuracy, and the ability to hit small targets at long range. It is usually paired with a telescopic sight, which helps the user see distant objects clearly and aim with greater precision.

The history of sniper rifles can be traced back several centuries, but their modern form developed during the nineteenth and early twentieth centuries. Early forms of long-range shooting appeared when firearms became more reliable and accurate. During the American Civil War and later the First World War, soldiers began using rifles with improved sights to target enemies at a distance. By the time of the Second World War, sniper rifles had become an important part of military strategy. Specially trained individuals, known as snipers, were assigned to use these rifles to observe and engage specific targets. Over time, improvements in materials, optics, and manufacturing techniques made sniper rifles even more accurate and effective.

A sniper rifle is built with several key features that support precision. The barrel, the part that guides the bullet, is carefully manufactured to ensure consistency and accuracy. The trigger system is designed to be smooth and predictable, allowing the shooter to fire without disturbing the aim. The stock, or body of the rifle, is shaped to provide stability and comfort. In addition, most sniper rifles are equipped with advanced optical sights that can magnify the target, making it easier to aim at long distances. Some rifles also include support systems such as bipods, which help keep the rifle steady during use.

The process of making a sniper rifle involves careful engineering and quality control. Manufacturers use strong, durable materials to ensure the rifle performs reliably under various conditions. Precision machining is used to create parts that fit together perfectly, reducing any variation that could affect accuracy. The optical systems are also carefully designed to provide clear and reliable views of distant targets. Because of the importance of accuracy, each rifle is often tested to ensure that it meets strict performance standards.

One of the main advantages of sniper rifles is their precision. They allow a trained user to target specific objects or individuals with high accuracy. This can reduce the need for widespread use of force and limit unintended damage. In military operations, sniper rifles can be used to observe and control situations from a distance, providing valuable information as well as targeted action. In law enforcement, they may be used when a precise response is needed to protect lives.

Another advantage is their range. Sniper rifles can be effective over distances much greater than standard firearms. This allows the user to remain at a safe distance while still observing and engaging targets. The ability to operate from a distance can reduce exposure to danger and provide a tactical advantage in certain situations. In addition, sniper rifles are often used for surveillance, as the optical equipment can help in observing areas without being detected.

However, the disadvantages and risks associated with sniper rifles are also important to consider. One major concern is their potential to cause harm to individuals. Because they are designed for precision, they can be used for targeted killing. This raises ethical questions, especially in situations where the use of force may not be clearly justified. The ability to strike from a distance can make it difficult for the target to respond or defend themselves.

Another issue is the potential for misuse. While trained professionals typically use sniper rifles, there is a risk that such weapons could be used by individuals outside of official control. This can increase the danger to public safety. The accuracy and range of these rifles mean that they can be used in ways that are difficult to prevent or respond to quickly.

The impact on human life can be serious. In conflict situations, sniper rifles can be used to create fear and uncertainty. The presence of a sniper can affect how people move and behave, as they may feel unsafe even in areas that are not directly under attack. This psychological effect can influence both soldiers and civilians, adding to the stress of the situation.

Human settlements can also be affected by the use of sniper rifles. In urban areas, their use can disrupt daily life, as people may avoid certain locations or activities due to safety concerns. In conflict zones, the presence of snipers can limit movement and access to essential services such as food, water, and medical care. This can make it difficult for communities to function normally and can slow down recovery efforts.

The ethical and legal aspects of sniper rifle use are often discussed in relation to rules of engagement and the protection of civilians. In military and law enforcement settings, there are guidelines that aim to ensure that force is used only when necessary and in a controlled manner. The precision of sniper rifles can support these guidelines, but it also requires careful judgment and responsibility from those who use them.

The cost of sniper rifles can vary depending on their design and features. High-quality rifles with advanced optics and materials can be expensive to produce and maintain. Training for users is also important, as effective use requires skill, practice, and an understanding of the equipment. This training includes not only technical skills but also awareness of safety and legal considerations.

It is also worth noting that the technology used in sniper rifles has connections to other fields. The development of precision engineering, optics, and materials has applications in sports shooting, hunting, and scientific research. This shows that the same knowledge can be used for both beneficial and harmful purposes, depending on how it is applied.

In conclusion, sniper rifles are high-precision firearms designed for accurate targeting over long distances. They have a long history of development and are used in military and law enforcement contexts for specific purposes. While they offer advantages such as precision and range, they also raise concerns about their potential for harm and misuse. Their impact on human life and communities can be significant, especially in conflict situations. Understanding how sniper rifles work and their effects is important to ensure they are used responsibly and within established guidelines.

18. Assault Rifles

Assault rifles are firearms designed to fire multiple rounds in a short period of time and to be used by a single person with relative ease. They are usually capable of both semi-automatic fire, where one round is fired for each pull of the trigger, and automatic or burst fire, where several rounds can be fired in quick succession with a single trigger action. Military forces around the world widely use these rifles and, in some countries, they are also present in civilian settings under regulated conditions. Their design balances power, portability, and rate of fire, making them effective in many situations but also raising concerns about misuse.

The development of assault rifles began in the early twentieth century, as military forces sought weapons that could deliver both range and rapid fire. During the Second World War, new types of rifles were developed that could fire intermediate cartridges, which are less powerful than traditional rifle ammunition but allow for better control and faster shooting. One of the earliest examples of this concept was introduced by Germany during the war. After that, other countries began to design their own versions, leading to well-known models such as the AK-47 and the M16. Over time, assault rifles became standard equipment for many armed forces because they offered flexibility in combat.

The design of an assault rifle includes several important parts. The barrel guides the bullet and affects accuracy. The firing mechanism controls how the weapon operates in semi-automatic or automatic mode. The magazine holds multiple rounds of ammunition, allowing the user to fire several shots before needing to reload. The stock and grip are shaped to make the rifle easier to hold and control. Modern assault rifles often include additional features such as adjustable sights, attachment rails, and lighter materials that reduce weight while maintaining strength.

The process of making an assault rifle involves advanced manufacturing techniques. Strong metals and durable materials are used to ensure that the weapon can handle repeated use. Precision machining is required to create parts that fit together correctly and function reliably. The assembly process must be carefully controlled to ensure safety and performance. Because of the potential risks, the production and distribution of such firearms are regulated by laws in many countries. These laws vary by region and are intended to regulate access and reduce misuse.

One of the main advantages of assault rifles is their versatility. They can be used in different modes depending on the situation, which makes them adaptable. For example, semi-automatic fire can be used for more controlled shooting, while automatic fire can be used when a higher volume of fire is needed. This flexibility makes them useful in military operations where conditions can change quickly. Another advantage is their relatively compact size compared to older rifles, which makes them easier to carry and use in different environments.

Assault rifles are also designed to be user-friendly. Their controls are generally straightforward, and they can be operated with training. This ease of use is one reason armed forces widely adopt them. In addition, their ability to fire multiple rounds without frequent reloads can be an advantage in situations requiring continuous action.

However, the disadvantages of assault rifles are significant and have led to ongoing debate. One major concern is their potential for misuse. Because they can fire rapidly and are relatively easy to handle, they can cause serious harm if used in violent situations. Incidents involving such firearms have raised questions about access and regulation, especially in civilian contexts. The ability to fire many rounds in a short time increases the risk of injury and loss of life.

Another issue is the impact on public safety. In areas where assault rifles are used improperly, there can be a higher risk of violence. This can affect communities by creating fear and reducing the sense of security.

People may feel unsafe in public spaces, which can affect daily life and social interactions. The presence of such weapons in civilian settings is often a topic of discussion among policymakers and the public.

The impact on human life is one of the most serious aspects. Assault rifles can cause severe injuries due to the speed and energy of the bullets they fire. In conflict situations, they contribute to casualties and destruction. In non-military settings, their misuse can lead to tragic events that affect individuals, families, and communities. The physical and emotional effects of such incidents can last for a long time.

Human settlements can also be affected by the presence and use of assault rifles. In conflict zones, they are often used in urban areas, which can damage buildings and infrastructure. This can disrupt services such as electricity, water supply, and transportation. In civilian areas, safety concerns can lead to changes in how people live, such as increased security measures or restrictions on movement. These modifications can influence the general quality of life.

The economic impact of assault rifles is another factor to consider. Governments spend resources on law enforcement, regulation, and public safety measures related to firearms. In areas affected by violence, there may be costs related to healthcare, property damage, and loss of productivity. These costs can add up over time and place a burden on society.

There are also ethical and legal considerations. The use of assault rifles in military operations is governed by rules that aim to limit harm and protect civilians. In civilian settings, laws are designed to control who can own and use such firearms. These laws vary widely and are often the subject of debate. Some people argue for stricter controls to reduce risk, while others emphasize the importance of legal access under certain conditions.

It is important to note that the technology used in assault rifles is part of a broader field of engineering and manufacturing. Advances in materials, design, and production have applications in many areas beyond firearms. This shows that the knowledge itself is not harmful, but its use must be carefully managed.

Efforts to address the risks associated with assault rifles include education, regulation, and enforcement of laws. Training programs for users can emphasize safety and responsibility. Law enforcement agencies work to prevent illegal use and to respond to incidents. Public discussions and policies aim to find a balance between different concerns, such as safety and rights.

In conclusion, assault rifles are firearms designed for rapid and flexible use, with a history that reflects changes in military needs and technology. They offer advantages such as versatility and ease of use, but they also pose serious risks when misused. Their impact on human life, communities, and society is significant, especially in cases of violence. Understanding their design, history, and effects is important for making informed decisions about their use and regulation. Many people believe that careful management and responsible use are essential to reduce harm and ensure safety.

19. Machine Guns

Machine guns are firearms designed to fire a continuous stream of bullets as long as the trigger is held down and ammunition is available. They are fully automatic, which means they do not require a separate trigger pull for each shot. This allows them to release a large number of rounds in a very short time. Machine guns are mainly used in military settings where sustained fire is needed to control areas, support troops, or defend positions. Their ability to deliver high firepower makes them among the most impactful weapons on the battlefield.

The history of machine guns dates back to the late nineteenth century, when inventors began seeking ways to increase the rate of fire of weapons. One of the earliest successful designs was created by Hiram Maxim in the 1880s. His machine gun used the energy from each fired shot to reload and fire the next round automatically. This was a major advancement because it allowed continuous fire without manual action between shots. Machine guns played a significant role in the First World War, where they were used to defend positions and caused heavy losses. Over time, their design improved, making them lighter, more reliable, and easier to operate. By the Second World War and beyond, machine guns became standard equipment for many armies.

A machine gun is built with several key parts that work together to allow automatic firing. The barrel directs the bullet and must be strong enough to handle repeated firing without damage. The action mechanism uses the energy from each shot to load the next round and prepare the weapon for the next shot. The feeding system supplies ammunition, often from a belt or a large magazine, allowing for continuous fire. The cooling system is also important, as firing many rounds quickly can generate significant heat. Some machine guns use air cooling, while others may use heavier barrels or other methods to manage heat.

The process of making a machine gun requires careful engineering and precise manufacturing. Tough materials are chosen to ensure longevity and safety. Parts must be made to exact measurements so that they fit together correctly and function reliably. Because machine guns operate at high speeds and under stress, quality control is very important. Governments usually control the production and distribution of such weapons due to their potential impact.

One of the main advantages of machine guns is their high rate of fire. They can fire hundreds of rounds per minute, allowing them to cover a large area with bullets. This can be useful in military situations where suppressing enemy movement is important. By creating a continuous stream of fire, machine guns can limit opponents' ability to move or act. This makes them valuable for defense and support roles.

Another advantage is their ability to provide sustained fire. Unlike weapons that require frequent reloading, machine guns can continue firing for longer periods, especially when supplied with ample ammunition. This can be important in situations where continuous action is needed. Machine guns are also used to support other troops, helping to protect them as they move or carry out tasks.

However, the disadvantages of machine guns are significant and must be carefully considered. One major issue is the level of destruction they can cause. The high number of rounds fired in a short time increases the risk of injury and loss of life. In conflict situations, machine guns have been responsible for large numbers of casualties. Their ability to affect a wide area means they can impact not only intended targets but also nearby targets.

Another concern is their use in populated areas. When machine guns are used in or near civilian environments, the risk to non-combatants increases. Buildings, vehicles, and other structures can be damaged, and people may be harmed even if they are not directly involved in the conflict. This raises important questions about how and where such weapons should be used.

The impact on human life is profound. Machine guns can cause severe injuries due to the speed and force of the bullets. In addition to physical harm, their presence can create fear and stress among those who are exposed to them. In war zones, the sound and effect of machine gun fire can influence behavior, limiting movement and affecting daily life.

Human settlements can also be affected by the use of machine guns. In areas where fighting takes place, infrastructure such as homes, roads, and public facilities may be damaged. This can disrupt essential services and make it difficult for communities to function. People may be forced to leave their homes, leading to displacement and long-term challenges. Recovery can take time and resources, especially in regions where damage is widespread.

The economic impact of machine guns is another factor. The production and maintenance of these weapons require investment. In addition, areas affected by their use may incur costs for rebuilding and supporting those affected. Healthcare systems may also bear the burden of treating injuries caused by such weapons. These costs can affect both local and national economies.

There are also ethical and legal considerations related to machine guns. Their use in military operations is governed by rules that aim to limit harm and protect civilians. These rules are part of international agreements that guide how weapons should be used in conflict. In civilian settings, laws often restrict access to fully automatic firearms due to their potential for misuse. These regulations vary by country and are often the subject of discussion and debate.

Another issue is the potential for misuse. While machine guns are typically used by trained personnel in controlled environments, there is always a risk that they could be used in harmful ways outside of these settings. This has led to strict controls in many places to prevent unauthorized access and to ensure that such weapons are used responsibly.

It is also important to recognize that the technology behind machine guns is part of a broader field of mechanical and materials engineering. Advances in these areas have applications in many industries, including manufacturing and transportation. This shows that the knowledge used to create such weapons can also be used for positive purposes.

Efforts to reduce the risks associated with machine guns include training, regulation, and international cooperation. Military personnel are trained to use these weapons responsibly and to follow rules that aim to protect civilians. Governments enforce laws to control access and prevent misuse. International agreements seek to limit the impact of weapons in conflict and promote safer practices.

In conclusion, machine guns are powerful firearms designed for sustained and rapid fire. They have played a significant role in the history of warfare and continue to be used in military operations. While they offer advantages such as high firepower and the ability to support troops, their potential for causing harm is considerable. Their impact on human life, communities, and the environment highlights the need for careful control and responsible use. Understanding how machine guns work and the effects they can have is important for making informed decisions about their role in society.

20. Rocket-Propelled Grenades (RPGs)

Rocket-propelled grenades, often called RPGs, are portable weapons that fire explosive projectiles using a rocket system. They are designed to be carried and operated by a single person, making them easy to move and use in various environments. An RPG system usually consists of a launcher, a tube-like device, and a rocket with an explosive warhead. When fired, the rocket travels toward the target and detonates on impact or after a short delay. These weapons are widely used in combat zones because they can be effective against vehicles, defensive positions, and buildings.

The history of rocket-propelled grenades dates back to the early twentieth century, when military forces began exploring ways to combine rockets with explosive payloads. During the Second World War, several countries developed early versions of these weapons to deal with armored vehicles. These early designs were not always reliable, but they showed that portable rocket weapons could be useful on the battlefield. After the war, technological advances led to more advanced and reliable systems. During the Cold War, RPGs became widely produced and distributed, especially in conflict zones. Over time, they became one of the most commonly used portable anti-armor weapons in the world.

An RPG works through a simple yet effective mechanism. The launcher is aimed at a target, and when the trigger is pressed, the rocket is fired from the tube. An initial charge first pushes out the rocket, and then its motor ignites, propelling it forward at high speed. The warhead at the front of the rocket is designed to explode upon reaching the target. Some warheads are shaped to focus the explosive force in a specific direction, which increases their ability to penetrate armor or solid structures. The design allows the weapon to be used in a wide range of situations, from open fields to urban environments.

The production of RPG systems involves both mechanical and chemical engineering. The launcher must be strong enough to handle repeated use while remaining lightweight for portability. The rocket itself contains fuel for propulsion and an explosive charge for impact. These components must be carefully designed and assembled to ensure safety and effectiveness. Because of their military use, the production and distribution of RPGs are regulated in many countries, although they have also appeared in areas with weaker regulation.

One of the main advantages of RPGs is their portability. Unlike larger weapons systems that require vehicles or heavy equipment, RPGs can be carried by a single person. This allows them to be used in a wide range of locations, including hard-to-access areas. Their relatively simple design also makes them easier to operate compared to more complex weapons. With training, users can learn to handle and fire RPGs effectively.

Another advantage is their effectiveness against certain targets. RPGs are often used against vehicles, especially lightly armored ones. They can also be used to damage buildings or defensive positions. This makes them useful in situations where other weapons may not be practical. Their ability to deliver explosive force quickly and directly is one reason they are widely used in conflicts.

However, the disadvantages and risks associated with RPGs are serious. One major concern is the level of destruction they can cause. The explosive impact of a rocket can result in significant damage to both targets and surrounding areas. In urban environments, this can lead to harm to civilians and damage to infrastructure. Buildings, roads, and public facilities may be affected, making it difficult for communities to function.

Another issue is their ease of use. While this can be seen as an advantage in controlled settings, it also increases the risk of misuse. Because RPGs are portable and relatively simple to operate, they can be used by individuals or groups outside of formal military structures.

This raises concerns about safety and security, especially in areas where conflicts are ongoing or where weapons are not tightly controlled.

The impact on human life can be severe. The explosion from an RPG can cause injuries or death to those near the point of impact. In addition to physical harm, the use of such weapons can create fear and uncertainty among people living in affected areas. The sound and effect of explosions can disrupt daily life and lead to long-term psychological stress.

Human settlements are particularly vulnerable to the effects of RPGs. In conflict zones, these weapons are often used in or near populated areas. This can damage homes, schools, hospitals, and other important structures. The destruction of infrastructure can interrupt essential services such as water supply, electricity, and healthcare. People may be forced to leave their homes, leading to displacement and social disruption.

The economic impact of RPG use is also significant. Repairing damaged buildings and infrastructure requires time and resources. Businesses may be affected if areas become unsafe or if facilities are destroyed. The cost of recovery can place a burden on communities and governments, especially in regions that are already facing challenges.

There are also environmental effects to consider. Explosions can release dust and debris into the air, affecting air quality. Damage to buildings and land can alter the local environment and create hazards, such as unstable structures or unexploded ordnance. These effects can take time to address and may require cleanup efforts.

From an ethical and legal perspective, the use of RPGs is subject to rules that aim to limit harm and protect civilians. In military operations, some guidelines govern how and where such weapons can be used. However, when these rules are not followed, the impact on civilians can be severe. This has led to discussions about the need for better control and regulation.

Another concern is the spread of RPG technology. Because these weapons have been widely produced and distributed, they can be found in many parts of the world. This increases the risk that they may be used in conflicts or by groups that do not follow established rules. Efforts to control the spread of such weapons are part of broader international security measures.

It is also important to recognize that the technology behind RPGs is related to broader developments in engineering and materials science. The principles of propulsion and explosive force are studied for various purposes, including peaceful applications. This highlights the importance of using scientific knowledge responsibly.

In conclusion, rocket-propelled grenades are portable weapons that use rocket propulsion to deliver explosive force to a target. They have a long history of development and are widely used in military contexts due to their effectiveness and ease of use. However, their potential to harm people, communities, and infrastructure makes them a serious concern. Their impact on human life and settlements can be significant, especially in areas where they are used in conflict. Understanding how RPGs work and their effects is important for promoting responsible use and reducing risks.

21. Flamethrowers

Flamethrowers are devices designed to project a stream of burning fuel toward a target, creating fire over a short to medium distance. They are known for producing intense heat and visible flames, which can ignite materials and cause severe burns. A typical flamethrower system includes a fuel tank, a mechanism to pressurize and expel the fuel, and an ignition system that ignites the fuel as it exits the device. Once released, the burning fuel can stick to surfaces and continue to burn, making the effect both immediate and lasting. These devices have been used in warfare and in certain controlled operations, such as clearing vegetation or obstacles, although their use is often limited due to safety concerns.

The history of flamethrowers dates back to ancient times in a very basic form, when people used fire as a weapon in different ways. However, modern flamethrowers were developed in the early twentieth century. One of the first widely recognized uses of flamethrowers was during the First World War. They were introduced as a means of attacking fortified positions, such as trenches and bunkers, where traditional weapons were less effective. Soldiers could direct flames into confined spaces, forcing opponents to leave their positions. Flamethrowers continued to be used in later conflicts, including the Second World War and subsequent wars. Over time, their design improved, but their use became more restricted due to the risks and ethical concerns associated with them.

A flamethrower works by storing a flammable liquid, usually a type of fuel, in a pressurized tank. When the operator activates the device, the fuel is forced out through a nozzle. At the same time, an ignition system at the front of the nozzle lights the fuel, creating a stream of fire. The flame's range depends on the device's design and the pressure used to deliver the fuel. The burning fuel can reach targets several meters away and can spread over surfaces, depending on the conditions. The heat produced is very high, which allows the flames to ignite materials quickly.

The construction of a flamethrower requires careful design to ensure it operates safely and effectively. The fuel tank must be strong enough to hold pressurized liquid without leaking. The delivery system must control the flow of fuel to ensure accurate delivery. The ignition system must be reliable to ensure that the fuel ignites at the right moment. Because of the risks involved, the development and use of flamethrowers are usually controlled by military or specialized organizations. Detailed information about their design is not widely shared.

One of the main advantages of flamethrowers is their ability to create intense heat over a targeted area. This makes them effective in situations where fire can clear obstacles or force opponents out of protected positions. The visual and psychological impact of flames can also be significant, as fire can create fear and confusion. In some controlled settings, flamethrowers have been used to clear vegetation or remove hazards, although such uses require strict safety measures.

However, the disadvantages of flamethrowers are serious and often outweigh their advantages. The most important concern is the level of harm they can cause to people. The flames produced by these devices can result in severe burns, which are often life-threatening and can lead to long-term physical and emotional suffering. Because the fuel can adhere to surfaces, including skin, the burning can persistlonger, increasing the severity of injuries.

Another major issue is the difficulty of controlling fire. Once a flame is released, it can spread beyond the intended target, especially in environments with flammable materials. This can lead to unintended damage to buildings, vegetation, and other structures. In urban or populated areas, the risk of fire spreading is particularly high, which can result in large-scale destruction.

The impact on human settlements can be significant. Fires caused by flamethrowers can destroy homes, infrastructure, and public facilities. This can leave people without shelter and disrupt essential services such as water supply, electricity, and transportation. The recovery process can be long and costly, as rebuilding requires resources and time. In addition, the presence of fire can create panic, making it harder for people to respond effectively to emergencies.

The environmental impact of flamethrowers is also important to consider. Fires can damage ecosystems by destroying plants and harming animals. The smoke produced can affect air quality and may have health effects on people who inhale it. In some cases, environmental effects can persist for a long time, especially if large areas are burned.

The psychological effect of flamethrowers is another serious concern. The sight and experience of intense fire can create fear and trauma among those who are exposed to it. Survivors of such events may experience long-term stress and anxiety. The use of fire as a weapon has a strong emotional impact on both individuals and communities.

From an ethical and legal perspective, the use of flamethrowers has been widely debated. Many people believe that weapons that cause severe suffering and are difficult to control should be restricted. International laws and agreements have been developed to limit the use of certain types of incendiary weapons, especially in areas where civilians are present. These rules aim to reduce harm and protect people during conflicts.

The economic impact of flamethrowers can also be significant. The destruction caused by fires can lead to high costs for rebuilding and recovery. Healthcare systems may face the burden of treating burn injuries, which often require specialized care. These costs can affect both local communities and larger systems, especially in areas where resources are limited.

It is also important to note that the science behind flamethrowers, such as the study of fuels and combustion, has many positive applications. This knowledge is used in energy production, manufacturing, and other industries that support modern life. The challenge is to ensure that such knowledge is used in ways that benefit society rather than cause harm.

Efforts to reduce the risks associated with flamethrowers include regulation, training, and the development of safer alternatives. In many places, their use is limited or controlled to prevent misuse. Awareness of fire dangers and proper safety measures is important for reducing accidents and harm.

In conclusion, flamethrowers are devices that project burning fuel to create intense heat and fire. They have been used in warfare and other operations for their ability to clear obstacles and engage targets with fire. However, their potential to cause severe harm, spread uncontrollable fires, and damage communities and the environment makes them highly dangerous. Their impact on human life and settlements highlights the need for careful control and responsible use. Understanding how flamethrowers work and the risks they pose is important for promoting safety and reducing harm.

22. Hand Grenades

Hand grenades are small explosive devices designed to be thrown by hand, usually in close-range situations. They are commonly used in military operations and sometimes by law enforcement units in controlled environments. A hand grenade is designed to explode after a short delay, giving the user time to throw it toward a target before it detonates. When it explodes, it releases energy and fragments that can cause damage within a limited area. Because of their compact size and ease of use, hand grenades have been widely used in different types of conflicts over many years.

The history of hand grenades dates back several centuries, with early versions taking simple forms. In the past, small containers filled with explosive materials were thrown by soldiers during battles. These early devices were not very reliable, but they showed the idea of using a small, portable explosive. During the First World War, hand grenades became more advanced and were used extensively in trench warfare. Soldiers needed a weapon that could be used in close quarters, and grenades provided a way to reach enemies behind cover. Over time, improvements in design and materials made grenades safer to handle and more effective in use. They continued to be used in later conflicts and remain part of modern military equipment.

A hand grenade typically consists of a metal or composite outer casing, an explosive charge inside, and a triggering mechanism. The casing is constructed to shatter into smaller fragments when the grenade detonates. These pieces, often called fragments, spread out in all directions and can cause injury or damage. The triggering mechanism usually includes a safety pin and a lever. When the user removes the pin and releases the lever, a delay fuse is activated. This fuse burns for a short period, often a few seconds, before setting off the main explosion. This delay allows the grenade to be thrown to a safe distance before it detonates.

The process of making hand grenades involves careful design and manufacturing. The materials used must be strong enough to contain the explosive until detonation, but also able to break apart in a controlled manner. The fuse system must be reliable to ensure the grenade explodes at the correct time. Safety features are also important to prevent accidental activation during handling and transport. Because of the risks involved, the production of hand grenades is controlled by governments and carried out under strict conditions.

One of the main advantages of hand grenades is their simplicity. They are easy to carry and can be used without complex equipment. This makes them useful in situations where other weapons may not be practical. Their small size allows them to be used in confined spaces, such as buildings or trenches. Another advantage is their ability to affect an area rather than a single point. The blast and shrapnel can reach targets behind cover, making grenades effective in close-combat situations.

Hand grenades are also versatile. Different types of grenades are designed for specific purposes. For example, fragmentation grenades are intended to cause damage through fragments, while others may produce smoke or light for signaling or distraction. This variety allows them to be used in different roles depending on the situation.

However, the disadvantages of hand grenades are significant. One major concern is their potential for causing harm beyond the intended target. The fragments from the explosion spread in all directions, meaning anyone within the blast radius can be affected. This makes them dangerous in areas where civilians or friendly forces are nearby. The limited control over where fragments travel increases the risk of unintended injury.

Another issue is the risk of accidents. Although grenades are designed with safety features, improper handling or malfunction can lead to unintended explosions. This can pose a danger not only to the user but also to others in the area. Training and careful handling are essential to reduce these risks, but they cannot be eliminated.

The impact on human life is serious. The explosion of a hand grenade can cause injuries such as wounds from fragments, burns, and damage from the blast itself. Such injuries can be critical and may demand urgent medical intervention. In addition to physical harm, the use of grenades can create fear and stress among those who are exposed to them, especially in conflict zones.

Human settlements can also be affected by the use of hand grenades. In urban areas, their use can damage buildings, vehicles, and infrastructure. Even a single explosion can disrupt daily life, especially if it occurs in a crowded place. Repeated use in conflict can cause significant damage and make it difficult for communities to function. This can affect access to essential services such as healthcare, education, and transportation.

The economic impact of hand grenades is often linked to the damage they cause. Repairing buildings and infrastructure requires resources and time. In areas affected by conflict, the cost of recovery can be high, especially when combined with other forms of damage. Healthcare systems may also face increased demand due to injuries caused by explosions.

There are also ethical and legal considerations related to the use of hand grenades. In military operations, some rules aim to limit harm and protect civilians. These rules guide how and where such weapons can be used. However, when these rules are not followed, the impact on civilians can be severe. The ease of use of grenades also raises concerns about their potential misuse outside of controlled environments.

Another important point is the spread of hand grenades. Because they are small and relatively simple, they can be transported easily. This increases the risk that they may be used by individuals or groups outside of official control. Efforts to regulate and control access to such weapons are part of broader security measures.

The environmental impact of hand grenades is usually limited compared to larger weapons, but it is still present. Explosions can create debris and damage the surrounding area. In areas where grenades are used frequently, the cumulative effect can result in visible damage and cleanup needs.

It is also worth noting that the technology behind hand grenades, such as the study of materials and controlled explosions, has applications in other fields. These include construction, mining, and engineering, where controlled use of energy can be beneficial. This shows that the same knowledge can have both positive and negative uses.

In conclusion, hand grenades are small, portable explosive devices designed for use in close-range situations. They have a long history of development and are used in military operations due to their simplicity and effectiveness. While they offer advantages such as ease of use and the ability to affect an area, they also pose serious risks. Their potential to harm people and damage communities makes them a significant concern. Understanding how hand grenades work and their effects is important for promoting responsible use and reducing harm.

23. Anti-Tank Missiles

Anti-tank missiles are weapons designed to destroy or disable armored vehicles such as tanks and other heavily protected military equipment. These missiles are guided, meaning they can be directed toward a target after launch. This guidance makes them more accurate than older weapons that relied only on direct aim. Anti-tank missiles are widely used in modern warfare because armored vehicles play a major role in military operations, and there is a need for weapons capable of effectively countering them. These missiles are used by trained personnel and are often part of infantry units, vehicle systems, or aircraft equipment.

The development of anti-tank missiles began during the twentieth century, especially after tanks became an important part of warfare. During the First World War, tanks were introduced to traverse difficult terrain and breach defenses. As tanks became stronger and more advanced, new weapons were needed to stop them. Early anti-tank weapons included rifles and simple rockets, but these had limited effectiveness against improved armor. During and after the Second World War, research led to the development of guided missile systems capable of tracking and hitting moving targets. By the Cold War period, anti-tank missiles had become more advanced, with better accuracy and stronger warheads. Today, they are a key part of many military forces.

An anti-tank missile system usually includes a launcher, a missile, and a guidance mechanism. The launcher is used to aim and fire the missile. The missile contains a propulsion system, guidance controls, and a warhead designed to penetrate armor. Once launched, the missile travels toward the target, guided by systems that may use wires, lasers, or other forms of communication. Some systems require the operator to keep the target in sight, while others can guide themselves after launch. The warhead is often designed to focus its energy to penetrate thick armor and damage the vehicle.

The production of anti-tank missiles involves advanced engineering and careful design. The missile must be able to travel accurately and remain stable during flight. The guidance system must respond quickly and correctly to ensure that the missile reaches its target. The warhead must be powerful enough to penetrate armor, a requirement that demands precise design and testing. Because of their complexity and potential impact, these weapons are developed and produced under strict government and defense organization control. Detailed information about their construction is not publicly shared.

One of the main advantages of anti-tank missiles is their precision. Because they are guided, they can be directed toward specific targets with high accuracy. This allows operators to focus on important targets and reduce unnecessary damage to surrounding areas. Another advantage is their effectiveness against armored vehicles. Tanks and similar vehicles are built to withstand many types of attacks, but anti-tank missiles are specifically designed to overcome this protection. This makes them an important tool in balancing military capabilities.

Another benefit is their flexibility. Anti-tank missile systems can be used from different platforms, including ground positions, vehicles, and aircraft. This allows them to be deployed in various environments and situations. Some systems are portable, allowing individual soldiers to carry and use them, while others are mounted on larger platforms for increased range and power. This adaptability makes them useful in many types of operations.

However, the disadvantages and risks of anti-tank missiles are significant. One major concern is their destructive power. The ability to penetrate armor means they can cause severe damage to vehicles and anyone inside. In addition, the explosion can affect nearby areas, especially in confined or populated environments. This increases the risk of harm to civilians and damage to infrastructure.

Another issue is the cost and complexity of these systems. Developing and maintaining anti-tank missiles requires advanced technology and resources. Training operators to use them effectively also takes time and effort. This can limit their availability and increase the overall cost of military operations. At the same time, the need to defend against such weapons can lead to further investment in protective systems, creating an ongoing cycle of development.

The impact on human life is serious. In conflict situations, anti-tank missiles contribute to casualties and destruction. The use of these weapons can lead to loss of life and injury, both among military personnel and, in some cases, civilians. The presence of such weapons can also increase tension and fear in affected areas, as people may feel unsafe due to the potential for sudden attacks.

Human settlements can be affected by the use of anti-tank missiles, especially in urban environments. If these weapons are used near buildings or infrastructure, the resulting damage can disrupt essential services. Roads, communication systems, and public facilities may be affected, making it difficult for communities to function. In some cases, people may be forced to leave their homes due to safety concerns, leading to displacement and long-term challenges.

The environmental impact of anti-tank missiles is generally limited compared to larger weapons, but it can still be important. Explosions can cause localized damage to land and structures. Debris and remnants may need to be cleared to restore affected areas. In regions where conflict continues for a long time, repeated use of such weapons can contribute to broader environmental damage.

From an ethical and legal perspective, the use of anti-tank missiles is governed by rules that aim to limit harm and protect civilians. These rules are part of international agreements that guide how weapons should be used in conflict. However, the effectiveness of these rules depends on how they are followed in practice. When used responsibly and within guidelines, these weapons are intended to target military objectives. When misused, they can cause harm beyond their intended purpose.

Another concern is the spread of anti-tank missile technology. As these systems become more advanced, there is a risk that groups outside of formal military control may acquire them. This can increase the potential for misuse and raise security concerns. Efforts to control the spread of such technology are part of broader international cooperation.

It is also important to note that the technology behind anti-tank missiles has connections to other areas, such as aerospace and engineering. The study of guidance systems, propulsion, and materials has applications in fields like space exploration and transportation. This shows that the same knowledge can be used for both beneficial and harmful purposes.

In conclusion, anti-tank missiles are guided weapons designed to destroy armored vehicles with high precision. They have developed over time in response to changes in military technology and have become an important part of modern warfare. While they offer advantages such as accuracy and effectiveness, their potential for destruction and impact on human life and communities make them a serious concern. Understanding how these weapons work and the effects they can have is important for promoting responsible use and reducing harm.

24. Surface-to-Air Missiles (SAMs)

Surface-to-air missiles, often called SAMs, are weapons designed to target and destroy aircraft while they are in the air. These aircraft can include fighter jets, bombers, helicopters, and sometimes even uncrewed aerial vehicles. SAM systems are usually part of a country's air defense network, which is built to protect important areas such as cities, military bases, and infrastructure from aerial attacks. These missiles are launched from the ground and guided toward a target with high accuracy, making them a powerful tool in modern warfare.

The development of surface-to-air missiles began in the mid-twentieth century, especially during and after the Second World War. As aircraft became faster and more capable, traditional anti-aircraft guns were no longer sufficient to defend against them. This led to the need for more advanced systems capable of tracking and intercepting targets at greater distances and at higher speeds. During the Cold War, both the United States and the Soviet Union invested heavily in SAM technology. Early systems used radar to detect and track aircraft, and missiles were guided toward their targets using various methods. Over time, these systems became more sophisticated, with improved accuracy, range, and reliability. Today, many countries maintain SAM systems as a key part of their defense strategy.

A surface-to-air missile system typically includes several components working together. There is a radar or sensor system that detects and tracks incoming aircraft. A control unit processes this information and decides when to launch a missile. The missile itself contains a propulsion system, guidance system, and warhead. Once launched, the missile travels toward the target, guided by signals from the ground system or by its own onboard sensors. Some missiles use radar guidance, while others may use infrared or other tracking methods. The goal is to ensure that the missile reaches the target and detonates close enough to cause damage or destruction.

The production of SAM systems involves advanced engineering and technology. The radar systems must be able to detect targets at long distances and distinguish them from other objects. The missiles must be designed to travel at high speeds and remain stable during flight. The guidance systems must respond quickly to changes in the target's position. Because of the complexity and importance of these systems, they are developed and produced under strict government and specialized organization control. Detailed technical information is usually kept confidential.

One of the main advantages of surface-to-air missiles is their effectiveness in defending against aerial threats. They can engage targets at long distances, often before the aircraft can reach its objective. This provides a strong defensive capability and helps protect important locations. Another advantage is their accuracy. Modern SAM systems are designed to track targets carefully and adjust their path during flight, increasing the likelihood of a successful interception.

SAM systems also act as a deterrent. The presence of such defenses can discourage potential attacks, as opposing forces know that their aircraft may be at risk. This can influence military planning and reduce the likelihood of certain types of operations. In addition, SAM systems can operate in various environments, including fixed positions and mobile platforms, adding flexibility to their use.

However, the disadvantages and risks of SAM systems are also important to consider. One major concern is the potential for unintended harm. If a missile is launched incorrectly or if the system misidentifies a target, it could strike an unintended aircraft. This is especially concerning in areas where military and civilian aircraft operate in the same airspace. Such incidents can lead to loss of life and create serious international tensions.

Another issue is the impact on aviation. The presence of SAM systems in conflict zones can make air travel more dangerous. Civilian aircraft may need to avoid certain areas, leading to longer routes and increased costs. In some cases, airspace may be closed entirely, affecting travel and trade. This shows how the use of such weapons can have effects beyond the immediate conflict.

The impact on human life can be significant. When SAM systems are used in warfare, they contribute to the overall level of conflict and risk. The destruction of aircraft can result in loss of life, including both military personnel and, in some cases, civilians. The fear of aerial attacks and the presence of defense systems can also affect the mental well-being of people living in affected areas.

The deployment of SAM systems can influence human settlements. In areas where these systems are installed, there may be increased military activity, which can change the environment and daily life. In conflict zones, the use of SAMs can lead to damage if missiles or debris fall to the ground. This can affect buildings, roads, and other infrastructure. The need for security measures may also limit access to certain areas.

The economic impact of SAM systems is another factor. Developing and maintaining these systems requires significant investment in technology, training, and infrastructure. In addition, disruptions to air travel can affect industries such as tourism and trade. The cost of responding to incidents, including investigations and recovery efforts, can also be high.

There are also ethical and legal considerations. The use of surface-to-air missiles is governed by international laws that aim to protect civilians and ensure that force is used appropriately. These laws require careful identification of targets and consideration of the risks involved. However, in practice, mistakes can happen, and the consequences can be serious. This highlights the importance of training, clear procedures, and accountability.

Another concern is the spread of SAM technology. As these systems become more advanced, there is a risk that groups outside of formal military control may acquire them. This could increase the danger to both military and civilian aircraft. Efforts to control the spread of such technology are part of international security measures.

It is also important to recognize that the technology used in SAM systems has applications in other fields. Radar and tracking systems are used in aviation, weather monitoring, and scientific research. Advances in propulsion and guidance have contributed to developments in space exploration and other industries. This shows that the same knowledge can be used for positive purposes when applied responsibly.

In conclusion, surface-to-air missiles are advanced weapons designed to defend against aerial threats by targeting aircraft with high accuracy. They have developed over many years and play a key role in modern defense systems. While they offer advantages such as effective protection and deterrence, they also pose risks of unintended harm and impacts on aviation. Their effects on human life, communities, and global stability highlight the need for careful use and strong regulation. Understanding how these systems work and the challenges they present is important for ensuring safety and promoting responsible use.

25. Ballistic Missiles

Ballistic missiles are weapons designed to travel long distances by following a curved path, often described as a high arc through the atmosphere and sometimes into space before returning toward their target. Unlike some other missiles that are powered throughout their flight, ballistic missiles use rocket engines mainly during the early stage of launch. After this initial phase, they continue moving along a path determined by speed, gravity, and direction. Because of their speed and range, they are considered among the most powerful delivery systems in modern warfare.

The development of ballistic missiles began during the early twentieth century, with major progress made during the Second World War. One of the first well-known examples was the V-2 rocket developed by Germany. This rocket demonstrated that it was possible to send a weapon over long distances using rocket propulsion. After the war, scientists and engineers in other countries studied and improved this technology. During the Cold War, ballistic missile development became a major focus for countries like the United States and the Soviet Union. Both sides worked to create missiles that could travel farther, carry heavier payloads, and reach targets more quickly. This period led to the creation of intercontinental ballistic missiles, or ICBMs, which can travel across continents.

A ballistic missile typically works in several stages. In the first stage, the missile is launched using powerful rocket engines that push it upward and forward. This stage provides the speed needed to reach a high altitude. In the second stage, often called the midcourse phase, the missile travels through space or the upper atmosphere without active propulsion. During this time, it follows a path shaped by its initial speed and direction. In the final stage, the warhead re-enters the Earth's atmosphere and descends rapidly toward the target. The warhead can carry different types of payloads, depending on the missile's purpose.

The construction of ballistic missiles involves advanced engineering and careful design. The rocket must be strong enough to handle extreme forces during launch and flight. The guidance system must ensure that the missile follows the correct path and reaches its intended target. This system may use sensors, computers, and sometimes satellite navigation to adjust the missile's direction. The warhead must also be designed to function correctly after traveling long distances and experiencing high speeds. Because of their complexity and potential impact, ballistic missiles are developed under strict control, and detailed information about their design is not widely shared.

One of the main advantages of ballistic missiles is their range. They can travel hundreds or even thousands of kilometers, allowing them to reach distant targets without relying on nearby bases. This makes them useful for long-range operations and strategic planning. Another advantage is their speed. Once launched, ballistic missiles travel very quickly, especially during the final phase of their flight. This reduces the time available for detection and response, making them difficult to intercept.

Ballistic missiles can also carry large payloads. This allows them to deliver a significant force to a target. In some cases, a single missile can carry multiple warheads, each aimed at a different target. This increases their effectiveness and makes them more challenging to defend against. Their ability to act as a deterrent is also important. The presence of such weapons can discourage attacks, as opponents may fear a strong response.

However, the disadvantages of ballistic missiles are serious and far-reaching. One major concern is the level of destruction they can cause. Depending on the payload, a ballistic missile can cause significant damage to a target area. This can include the destruction of buildings and infrastructure, as well as loss of life. The scale of impact can be large, especially if the missile carries a powerful warhead.

Another issue is the difficulty of defense. Because ballistic missiles travel at high speeds and follow a high arc, intercepting them is challenging. Defense systems must detect the missile quickly and respond promptly. Even with advanced technology, it is not always possible to stop every missile. This creates a risk for areas that may be targeted.

The impact on human life can be severe. In the event of a missile strike, people in the affected area may face immediate danger as well as long-term challenges. Injuries, loss of life, and damage to homes and services can disrupt communities. The fear of such attacks can also affect mental well-being, creating stress and uncertainty.

Human settlements can be deeply affected by the use of ballistic missiles. Cities and towns may experience damage to buildings, roads, and essential systems such as water supply and electricity. Recovery can take a long time and require significant resources. In some cases, people may be forced to leave their homes, leading to displacement and social disruption.

The economic impact of ballistic missiles is also important. Developing and maintaining these systems requires large financial investments. In addition, the damage caused by their use can lead to high costs for rebuilding and recovery. Disruptions to trade, transportation, and business can affect economies at both local and national levels.

There are also environmental concerns. Explosions can release debris and harmful substances into the environment. Depending on the type of payload, there may be long-term effects on soil, water, and air. These effects can affect agriculture and natural ecosystems, compounding the challenges faced by affected areas.

From an ethical and legal perspective, the use of ballistic missiles is governed by international agreements and rules that aim to limit harm and protect civilians. These rules emphasize the importance of targeting only military objectives and avoiding unnecessary damage. However, the large scale of these weapons means that their use often raises serious concerns about their impact on people and the environment.

Another issue is the spread of ballistic missile technology. As more countries develop or acquire such systems, the risk of conflict and escalation may increase. Efforts to control the spread of missile technology are part of international security measures. These efforts aim to reduce the chances of misuse and promote stability.

It is also important to note that the technology used in ballistic missiles has connections to other fields, such as space exploration. The same basic principles of rocket propulsion are used to launch satellites and spacecraft. This shows that scientific knowledge can have both positive and negative uses, depending on how it is applied.

In conclusion, ballistic missiles are powerful weapons that travel long distances along a high-arc trajectory. They have developed over many years and play a significant role in modern military strategy. While they offer advantages such as range, speed, and deterrence, their potential for destruction and impact on human life and society make them a serious concern. Understanding how these weapons work and the effects they can have is important for promoting responsible use and reducing risks.

26. Submarine-Launched Missiles

Submarine-launched missiles are weapons that are fired from submarines while they are underwater. These missiles are designed to travel long distances and strike targets on land or at sea. They are an important part of modern military systems because they can be launched from hidden locations at sea. This ability to remain concealed makes them difficult to detect and track, which gives them a unique role in defense and strategic planning. Many of these missiles are capable of carrying powerful warheads, including nuclear ones, as well as conventional payloads.

The development of submarine-launched missiles began during the mid-twentieth century, especially during the Cold War. As countries developed ballistic missile technology, they also looked for ways to make these weapons more secure and less vulnerable to attack. Submarines provided a solution because they could move quietly underwater and remain hidden for long periods. Early systems were limited in range and required submarines to surface before launching. Over time, technological advances allowed missiles to be launched while the submarine remained underwater. This increased both safety and effectiveness. By the later stages of the Cold War, submarine-launched missiles had become a key part of military strategy for several countries.

A submarine-launched missile system includes the submarine itself, the missile, and the launch mechanism. The submarine is designed to carry and protect the missiles while remaining undetected. It uses quiet propulsion systems and special materials to reduce noise and avoid detection. The missiles are stored in vertical tubes inside the submarine. When a launch is ordered, the missile is pushed out of the tube using compressed gas or another system. Once it reaches the surface, its rocket engine ignites, and it continues its flight toward the target. Depending on the type, the missile may follow a ballistic path or use other guidance methods.

The construction of submarine-launched missiles involves advanced engineering and strict safety measures. The missile must be able to withstand pressure changes as it moves from underwater to the air. Its guidance system must be accurate enough to reach distant targets. The submarine must be designed to support long missions, often staying underwater for extended periods. These systems require careful planning, testing, and maintenance. Because of their complexity and importance, they are developed under strict control, and detailed information is not publicly shared.

One of the main advantages of submarine-launched missiles is their stealth. Submarines can operate quietly and remain hidden in large areas of the ocean. This makes it difficult for opponents to locate and track them. As a result, the missiles they carry can be launched from unexpected locations. This provides a strong strategic advantage and makes these systems an important part of deterrence. The idea of deterrence is that the presence of such weapons can discourage attacks, as opponents know that a response could come from hidden sources.

Another advantage is their survivability. Because submarines can move and hide, they are less vulnerable to attack compared to fixed launch sites. This means that even if a country's land-based systems are damaged, its submarine-based systems may still be able to respond. This ability to ensure a second response has been a key factor in maintaining balance between powerful nations. In addition, submarine-launched missiles often have long ranges, allowing them to reach targets far from their launch points.

However, the disadvantages and risks associated with these weapons are significant. One major concern is the level of destruction they can cause, especially when equipped with nuclear warheads. The impact of such weapons can be devastating, affecting large areas and causing long-term damage to people and the environment. Even when carrying conventional payloads, their ability to strike from hidden positions increases the risk of sudden, unexpected attacks.

Another issue is the difficulty of detection and response. While stealth is an advantage for those who possess these systems, it also creates challenges for defense. Opponents may find it hard to locate submarines and prepare for possible attacks. This can heighten tension and uncertainty, as countries may feel threatened by weapons that are difficult to track.

The impact on human life can be severe. In the event of an attack, the effects can include loss of life, injuries, and damage to infrastructure. The fear of such attacks can also affect people's sense of security. Living under the possibility of hidden threats can create stress and uncertainty, especially during times of conflict.

Submarine-launched missiles can have a serious effect on human settlements. Cities and towns targeted by such weapons may experience significant damage to buildings, roads, and essential services. Recovery from such damage can take a long time and require substantial resources. In some cases, people may be forced to leave their homes, leading to displacement and long-term challenges for communities.

The environmental impact is also an important consideration. Explosions, especially those involving powerful warheads, can affect land, water, and air. This can lead to contamination and damage to ecosystems. The long-term effects may include reduced agricultural productivity and harm to wildlife. These environmental changes can add to the difficulties faced by affected regions.

From an economic standpoint, the cost of developing and maintaining submarine-launched missile systems is very high. Building submarines, designing missiles, and training personnel require significant investment. At the same time, the damage caused by their use can lead to high costs for rebuilding and recovery. This raises questions about the allocation of resources and the balance between security and development.

There are also ethical and legal concerns. The use of such powerful weapons raises questions about the limits of force and the protection of civilians. International agreements aim to control the spread and use of nuclear weapons, including those carried by submarines. These agreements encourage countries to reduce their arsenals and promote peaceful solutions. However, progress has been gradual, and these weapons remain part of global security systems.

Another important issue is the risk of accidents or misunderstandings. Submarine operations involve complex systems and communication. Any error in judgment or technical failure could lead to serious consequences. In addition, misunderstandings between countries could increase the risk of conflict. This highlights the need for clear communication and strong safety measures.

It is also worth noting that the technology used in submarine-launched missiles has connections to other fields. Advances in underwater navigation, propulsion, and materials have applications in scientific research and exploration. For example, similar technologies are used in ocean studies and underwater vehicles. This shows that the same knowledge can be used for both beneficial and harmful purposes.

In conclusion, submarine-launched missiles are powerful weapons that combine advanced technology with the ability to operate from hidden underwater locations. They play an important role in modern military strategy, offering advantages such as stealth, range, and survivability. However, their potential for destruction and the challenges they pose to global stability make them a serious concern. Their impact on human life, communities, and the environment can be significant, especially if used. Understanding these systems and their effects is important for promoting responsible use, reducing risks, and encouraging efforts toward peace and cooperation.

27. Tactical Nuclear Weapons

Tactical nuclear weapons are a type of nuclear weapon designed for use in limited or specific battlefield situations rather than for large-scale, long-range attacks. They are often described as smaller in size and yield than strategic nuclear weapons, but this does not mean they are harmless. Even at a lower scale, they can cause serious destruction, produce dangerous radiation, and have lasting effects on people and the environment. These weapons are intended to target military positions, equipment, or specific areas rather than entire cities, but their impact can still extend beyond the intended target.

The development of tactical nuclear weapons began during the early years of the Cold War. After the use of atomic bombs in the Second World War, countries with nuclear capabilities started to explore different ways to use nuclear technology in military planning. Strategic nuclear weapons were designed for large targets and long distances, but there was also interest in weapons that could be used in smaller, more controlled situations on the battlefield. This led to the development of tactical nuclear systems, including bombs, artillery shells, and short-range missiles. These weapons were meant to provide an option for limited use, although the idea of "limited" nuclear use has always been a subject of debate.

Tactical nuclear weapons work on the same basic principles as other nuclear devices. They release a large amount of energy through nuclear reactions, either fission or a combination of fission and fusion. This release of energy produces a powerful blast, intense heat, and radiation. The size of the explosion depends on the design and purpose of the weapon. Tactical weapons are generally designed to have a lower yield than strategic ones, meaning their explosive power is lower. However, even a lower-yield nuclear explosion can cause significant damage within its range.

The creation of these weapons involves advanced scientific and engineering processes. Highly specialized materials are used, and the design must ensure that the nuclear reaction occurs in a controlled way at the moment of detonation. The systems used to deliver these weapons, such as short-range missiles or aircraft, must also be carefully designed to ensure accuracy and reliability. Because of the extreme risks and the sensitive nature of nuclear technology, the development and handling of these weapons are tightly controlled by governments. Detailed information about their construction is not publicly shared.

One of the main advantages of tactical nuclear weapons, from a military point of view, is their ability to deliver powerful effects on a specific target. They can be used to destroy heavily defended positions, large concentrations of equipment, or critical infrastructure. Because they are designed for shorter ranges and smaller targets, they may be seen as more flexible than larger nuclear weapons. This flexibility has led some to view them as a tool for limited use in certain situations.

Another advantage is their role in deterrence. The presence of tactical nuclear weapons can act as a warning to potential opponents. The idea is that the threat of their use may discourage attacks or aggressive actions. This concept has been part of military strategy for many years, especially during periods of high tension between powerful nations.

However, the disadvantages and risks of tactical nuclear weapons are very serious. One of the most important concerns is the level of destruction they can cause, even at a smaller scale. The blast from a nuclear explosion can destroy buildings and infrastructure, while the heat can cause fires and severe burns. The radiation released can affect people over both short and long periods, leading to illness and other health problems. These effects are not limited to the immediate area and can spread depending on conditions such as wind and geography.

Another major issue is the risk of escalation. Using a tactical nuclear weapon, even in a small-scale scenario, might escalate into a larger conflict. Once nuclear weapons are used, it may be difficult to prevent further use, which could result in larger and more destructive exchanges. This risk makes the use of such weapons highly controversial and dangerous.

The impact on human life is profound. People exposed to a nuclear explosion may suffer from immediate injuries as well as long-term health effects due to radiation. These effects can include serious illnesses that develop over time. In addition to physical harm, the psychological impact can be significant. The fear and trauma associated with nuclear events can affect individuals and communities for many years.

Human settlements can be heavily affected by tactical nuclear weapons. Even though they are intended for specific targets, the effects of the explosion and radiation can spread beyond the intended area. Buildings, roads, and essential services such as water and electricity can be damaged or destroyed. This can make it difficult for people to continue living in affected areas. In some cases, regions may become unsafe for extended periods due to radiation, leading to population displacement.

The environmental impact is another serious concern. Nuclear explosions can release radioactive materials into the air, soil, and water. This contamination can affect plants, animals, and ecosystems. It can also make land unsuitable for agriculture and affect food supplies. The long-term nature of radioactive contamination means that the effects can last for many years, creating ongoing challenges for recovery.

From an economic perspective, the use of tactical nuclear weapons can lead to high costs. These include the cost of immediate response, medical care, cleanup, and rebuilding. The disruption of economic activity can also have broader effects, affecting trade, employment, and overall stability. The financial burden can be especially heavy for areas that are already facing challenges.

There are also ethical and legal issues associated with these weapons. Many people question whether any use of nuclear weapons can be justified, given their impact on human life and the environment. International agreements aim to limit the spread and use of nuclear weapons, and there are ongoing efforts to reduce nuclear arsenals. However, tactical nuclear weapons remain part of military strategies in some countries, which continues to raise concerns.

Another important point is the risk of accidents or miscalculations. The systems used to manage nuclear weapons are complex, and any error could have serious consequences. Clear communication and strict safety measures are essential to reduce these risks. Even so, the possibility of unintended events cannot be completely removed.

It is also worth noting that the scientific knowledge used in nuclear technology has many positive applications. Nuclear energy is used to produce electricity, and research in this field has contributed to advances in medicine and science. This shows that the same knowledge can be used for both beneficial and harmful purposes, depending on how it is applied.

In conclusion, tactical nuclear weapons are smaller nuclear devices designed for use in specific battlefield situations. While they are considered more limited than larger nuclear weapons, their impact is still severe. They can cause destruction, radiation effects, and long-term harm to people and the environment. Their advantages in terms of flexibility and deterrence are outweighed by the serious risks they pose. Understanding these weapons and their consequences is important for promoting responsible decision-making and reducing the dangers they present to humanity and human civilization.

28. Poison Gas (Chlorine, Mustard Gas)

Poison gas, such as chlorine and mustard gas, is a type of chemical weapon that harms people by spreading toxic substances through the air. These gases can enter the body through breathing or contact with the skin and eyes. Once exposed, a person may experience serious health effects, including difficulty breathing, severe irritation, and long-term injury. Because these gases can spread over wide areas and affect many people at once, they are considered among the most harmful weapons developed in modern history.

The use of poison gas became widely known during the First World War. Although some forms of chemical warfare existed earlier, the large-scale use of gases like chlorine marked a major change in how wars were fought. In 1915, chlorine gas was released on the battlefield, creating a greenish cloud that moved with the wind toward enemy positions—soldiers who were exposed suffered from choking and lung damage. Soon after, other gases were developed, including mustard gas, which caused severe burns to the skin and eyes and could remain active in the environment for a long time. These weapons caused widespread suffering and led to strong reactions from the international community. Chlorine gas is widely used in industry for applications such as water treatment and manufacturing. However, when used as a weapon, it becomes dangerous because it reacts with moisture in the body to form acids that damage tissues, especially in the lungs. Mustard gas, on the other hand, is a chemical compound that affects the skin, eyes, and respiratory system. It does not always cause immediate pain, which makes it particularly harmful, as people may not realize they have been exposed until symptoms appear later. Both types of gas can spread through the air, carried by wind, making it difficult to control where they go.

The production of these gases involves chemical processes that are carefully controlled in industrial settings. In a military context, the gases are stored in containers and released to spread over a target area. This can be done through shells, bombs, or other delivery systems. Because of the dangers involved, the production and storage of such substances require strict safety measures. Today, international agreements place strong restrictions on the development and use of chemical weapons, and many countries have committed to eliminating their stockpiles.

One of the reasons poison gas was used in the past is its ability to affect large areas. Unlike weapons that target a specific point, gas can spread and reach many people at once. This made it a powerful tool for disrupting enemy forces and creating fear. Another factor was its psychological impact. The sight of a gas cloud and the knowledge of its effects could cause panic and confusion, making it harder for people to respond effectively.

However, the disadvantages and dangers of poison gas are very serious. One of the main issues is the suffering it causes. Exposure can lead to painful injuries, difficulty breathing, and long-term health problems. Mustard gas, for example, can cause severe skin burns and damage to the eyes, sometimes leading to blindness. Chlorine gas can damage the lungs, making it hard for a person to breathe. These effects can last long after the initial exposure and may require ongoing medical care.

Another major concern is the lack of control. Once poison gas is released, it can spread unpredictably depending on wind and weather conditions. This means it can affect not only the intended target but also nearby individuals, including civilians. In some cases, the gas can travel back toward those who released it, creating danger for both sides. This unpredictability makes it a highly risky weapon.

The impact on human life is severe and often long-lasting. Many people exposed to poison gas during the First World War suffered from health problems for the rest of their lives. Some developed chronic breathing issues, while others experienced lasting damage to their skin and eyes.

The fear and trauma associated with gas attacks also had a strong psychological effect on those who lived through them.

Human settlements can be deeply affected by the use of poison gas. In areas where gas is released, people may be forced to leave their homes to avoid exposure. Buildings and infrastructure may not be physically destroyed, but the area can become unsafe for a period of time. This can disrupt daily life and make it difficult for communities to function. In some cases, contamination may require cleanup efforts before people can return safely.

The environmental impact is also important to consider. Chemical substances released into the air can settle on the ground, affecting soil and water. This can harm plants and animals and may affect agriculture. The extent of the damage depends on the type of gas and the conditions in the area. While some chemicals break down over time, others may have longer-lasting effects.

From an ethical point of view, the use of poison gas has been widely condemned. The suffering it causes and the difficulty in controlling its spread have led to strong opposition. After the First World War, international agreements were created to limit and eventually ban the use of chemical weapons. These agreements reflect a global effort to prevent the use of such harmful substances in conflict.

The economic impact of poison gas can also be significant. Treating those who are affected requires medical resources, and long-term health problems can place a burden on healthcare systems. If areas are contaminated, there may be costs related to cleanup and restoration. In addition, the disruption of communities and businesses can affect local economies.

It is important to note that the chemicals used in poison gas have legitimate industrial uses when handled safely. Chlorine, for example, is important for water purification and helps make drinking water safe. This shows that the same substances can be beneficial or harmful depending on how they are used. The challenge is to ensure they are used responsibly, not as weapons.

In conclusion, poison gases such as chlorine and mustard gas are dangerous forms of chemical weapons that spread through the air and cause serious harm to people and the environment. Its use during the First World War revealed its devastating effects and led to strong efforts to prevent its use in the future. While it has certain characteristics that made it attractive as a weapon in the past, its disadvantages and impact on humanity are severe. The suffering it causes, the difficulty in controlling it, and its long-term effects make it one of the most concerning weapons ever developed. Understanding their history and consequences is important to ensure that such weapons are not used again and that efforts toward peace and safety continue.

29. Laser Weapons

Laser weapons are systems that use focused beams of light energy to affect a target. The term "laser" means light amplification through stimulated emission of radiation. In simple terms, a laser produces a highly concentrated beam of light that travels in a straight line and delivers energy to a small area. When used as a weapon, this energy can heat, damage, or turn off objects such as equipment, sensors, or, in some cases, people. Unlike traditional weapons that rely on physical projectiles or explosives, laser weapons use energy that travels at the speed of light, making their effects almost immediate.

The development of laser weapons is linked to advances in physics and technology during the twentieth century. The first working laser was demonstrated in 1960, and soon after, scientists began to explore its potential uses. Early research focused on communication, medicine, and industrial applications. However, military organizations also saw the potential of lasers as weapons. During the Cold War, research programs in several countries studied how high-energy lasers could be used for defense, including the possibility of intercepting missiles. Over time, improvements in power sources, cooling systems, and materials enabled the development of more practical laser systems. Today, laser weapons are still being refined and are used in limited roles, with ongoing research to expand their capabilities.

Laser weapons work by generating a highly focused, directed beam of light at a target. The system includes a power source, a laser generator, and a focusing mechanism. The power source provides the energy needed to produce the laser beam. The generator creates the light, and the focusing system directs it toward the target. When the beam hits the target, it transfers energy as heat. This heat can damage or destroy materials, turn off electronic systems, or interfere with sensors. The effectiveness of the laser depends on factors such as its power, the distance to the target, and environmental conditions.

The creation of laser weapons involves advanced engineering and scientific knowledge. The system must be able to produce a stable, powerful beam while managing heat and energy consumption. Cooling systems are important because generating high-energy lasers produces a lot of heat. The components must be carefully designed to ensure that the beam remains focused and accurate over distance. Because of the complexity and cost, these systems are developed under strict control, and detailed technical information is not widely available.

One of the main advantages of laser weapons is their precision. The beam can be directed to a very specific point, allowing for targeted action. This can reduce unintended damage to surrounding areas. Another advantage is speed. Since the laser beam travels at the speed of light, there is almost no delay between firing and impact. This makes it useful for engaging fast-moving targets, such as drones or missiles.

Laser weapons also have a continuous-supply advantage, as long as power is available. Unlike traditional weapons that rely on ammunition, a laser system can keep firing as long as it has energy. This can reduce the need for resupply and allow for extended operation. In addition, laser systems can be quieter than conventional weapons, potentially providing a tactical advantage in certain situations.

However, the disadvantages and limitations of laser weapons are also important to consider. One major challenge is the need for a strong and reliable power source. High-energy lasers require significant amounts of energy, which can limit their use in some environments. Managing heat is another issue, as the system can overheat if used continuously without proper cooling.

Environmental conditions can also affect the performance of laser weapons. Factors such as fog, rain, dust, or smoke can reduce the beam's strength or scatter the light, making it less effective. This means that their performance can vary depending on the situation. In addition, the range of a laser weapon may be limited compared to some other systems, especially in less favorable conditions.

The impact on human life can be serious if laser weapons are used against people. Focused energy can cause burns or eye damage, potentially leading to permanent injury. Even indirect exposure can be harmful, depending on the beam's strength. Because of these risks, there are international discussions about the use of lasers, especially those designed to cause blindness.

Human settlements can also be affected by the use of laser weapons, although their impact is different from that of explosive weapons. Instead of causing widespread destruction, lasers are more likely to target specific systems or structures. For example, they may be used to turn off equipment or damage critical components. While this can reduce large-scale damage, it can still disrupt services and affect daily life.

The economic impact of laser weapons is linked to their development and use. Building and maintaining these systems requires investment in research, materials, and training. At the same time, their ability to shut down equipment without destroying entire structures may reduce repair costs in certain situations. However, the overall financial impact depends on how they are used and the scale of their deployment.

There are also ethical and legal considerations. The use of laser weapons raises questions about safety and the protection of people. International agreements have been created to limit certain uses, such as lasers designed to cause permanent blindness. These rules aim to ensure that new technologies are used in ways that reduce harm and respect human rights.

Another important point is the potential for misuse. As technology advances, there is a risk that laser systems could be used in uncontrolled or unregulated ways. This highlights the need for clear guidelines and cooperation between countries to manage the risks associated with such weapons.

It is also worth noting that laser technology has many positive applications. Lasers are widely used in medicine, including surgeries and treatments. They are also used in communication, manufacturing, and scientific research. This shows that the same technology can be used for beneficial purposes when applied responsibly.

In conclusion, laser weapons are a modern development that uses focused light energy to affect targets with high precision and speed. They offer advantages such as accuracy, rapid response, and reduced reliance on traditional ammunition. However, they also face challenges related to power, environmental conditions, and potential risks to human safety. Their impact on society depends on how they are developed and used. Understanding these systems and their effects is important to ensure they are applied in a responsible, controlled manner, with attention to safety, ethics, and long-term consequences.

30. Railguns

Railguns are a type of advanced weapon that uses electromagnetic force instead of traditional explosives or chemical propellants to launch a projectile at very high speed. Unlike conventional guns that rely on burning fuel to push a bullet or shell out of a barrel, a railgun uses electricity to create powerful magnetic forces. These forces accelerate a solid projectile along a set of metal rails and send it toward a target at extremely high velocity. Because of this unique method, railguns are considered part of modern experimental and developing military technology.

The idea behind railguns comes from the study of electricity and magnetism, which has been understood for many years. Scientists learned that electric currents can produce magnetic fields and that these fields can create force. By the twentieth century, researchers began exploring how these principles could be used to move objects at high speeds. Early experiments showed that it was possible to launch projectiles using electromagnetic force, but the available power and materials limited the technology. In recent decades, advances in energy storage, electronics, and materials science have enabled the development of more practical railgun systems. Several countries have conducted research and testing, although the technology is still not widely used in active military operations.

A railgun operates with two parallel metal rails connected to a high-power source. A conductive projectile or armature is located between the rails. When a large electric current flows through the system, it creates a magnetic field around the rails. This field interacts with the current in the projectile, producing a force that propels it forward at very high speed. The projectile continues to accelerate along the rails until it leaves the barrel and travels toward the target. Because no explosive charge is used, the projectile relies on its speed and mass to cause damage.

The construction of a railgun requires advanced engineering and materials that can handle extreme conditions. The rails must be strong enough to carry very high electrical currents without melting or breaking. The system must include a powerful energy source, such as capacitors or other storage devices, to provide the necessary electricity in a short time. Cooling systems are also important, as the process generates a large amount of heat. The design must ensure that the projectile moves smoothly and that the system remains stable during operation. Because of these challenges, railguns are still in the testing and development stage in many places.

One of the main advantages of railguns is their extremely high speed. Projectiles fired from a railgun can travel much faster than those from traditional guns. This high speed increases the range and reduces the time it takes to reach the target. It also makes the projectile more difficult to intercept, as there is very little time to react. Another advantage is the absence of explosive material in the projectile. This can reduce the risk of accidental explosions during storage and transport.

Railguns also offer the potential for long-range accuracy. Because the projectile travels at such high speed, it can cover large distances while maintaining a relatively straight path. This can make it effective for targeting distant objects. In addition, using electricity instead of chemical propellants may allow repeated firing without the need for traditional ammunition, provided sufficient power is available.

However, the disadvantages and challenges of railguns are significant. One major issue is the need for a very powerful energy source. Generating the required amount of electricity in a short time is difficult and requires large, complex systems. This limits where railguns can be used, as not all platforms can support such power demands. Managing heat is another challenge, as the high currents produce intense heat that can damage the rails and other components.

The wear and tear on the system is also a concern. The extreme forces involved can cause the rails to degrade over time, requiring maintenance or replacement. This can increase the cost and complexity of using railguns. In addition, while the technology shows promise, it is still under development, and many technical challenges remain before it can be widely deployed.

The impact on human life can be serious if railguns are used in conflict. The high-speed projectiles can cause significant damage upon impact due to their kinetic energy. Even without explosives, the force of impact can destroy structures and harm people. The speed and power of these weapons mean there may be little warning before they strike, increasing the risk to those in the affected area.

Human settlements can also be affected by the use of railguns. In urban or populated areas, the impact of a high-speed projectile can damage buildings and infrastructure. Critical services including transportation, communication, and utilities could experience interruptions. Recovery from such damage can take time and resources, especially if the affected area is large.

The environmental impact of railguns is generally lower in terms of chemical pollution, since they do not rely on explosives or propellants. However, the physical damage caused by high-speed impacts can still affect land and structures. In addition, the energy required to operate railguns may have indirect environmental effects, depending on how that energy is produced.

From an economic perspective, the development and maintenance of railgun systems require significant investment. Research, testing, and production involve advanced technology and skilled personnel. While there is potential to reduce ammunition costs, the overall expense of building and supporting these systems remains high. This can limit their use to countries with strong technological and financial resources.

There are also ethical and strategic considerations. The introduction of new types of weapons can change the balance of power and influence how conflicts are managed. The speed and effectiveness of railguns may raise concerns about escalation and the ability to respond to threats. As with other advanced technologies, there is a need for clear rules and agreements to ensure responsible use.

Another concern is the potential spread of the technology. As research continues, railgun systems or similar technologies could become more widely available. This highlights the importance of monitoring and regulation to prevent misuse and maintain stability.

It is also important to recognize that the science behind railguns has positive applications. The study of electromagnetic forces and high-speed motion can contribute to advances in transportation, energy systems, and scientific research. For example, similar principles are used in experimental systems to launch objects into space or to improve industrial processes. This shows that the knowledge involved can benefit society when used responsibly.

In conclusion, railguns are advanced weapons that use electromagnetic force to launch projectiles at extremely high speeds. They represent a significant step in the development of modern military technology, offering advantages such as speed, range, and the use of non-explosive projectiles. However, they also face challenges related to power requirements, heat management, and system durability. Their potential impact on human life and infrastructure makes them a subject of concern, especially as the technology continues to develop. Understanding how railguns work and the issues they raise is important to ensure they are used responsibly and that their development is guided by careful consideration of safety, ethics, and long-term effects.

31. Hypersonic Missiles

Hypersonic missiles are advanced weapons that travel at speeds greater than five times the speed of sound, which is often described as Mach 5 or higher. At these speeds, a missile can move so quickly that it becomes very difficult to detect, track, and intercept. Unlike traditional missiles that follow more predictable paths, hypersonic missiles can change direction mid-flight, making them harder to defend against. Because of these features, they are considered one of the most significant developments in modern military technology.

The concept of hypersonic flight is based on scientific research into high-speed aerodynamics and propulsion. Scientists have studied how objects behave at very high speeds for many decades. Early research began in the mid-twentieth century, when countries started exploring rocket and jet technologies. During the Cold War, both the United States and the Soviet Union conducted experiments with high-speed vehicles, including hypersonic vehicles. However, the technology was difficult to develop due to challenges such as extreme heat, material strength, and control at high velocity. In recent years, advances in materials, computing, and engineering have enabled the development of more practical hypersonic systems. Today, several major countries are actively developing and testing these weapons.

Hypersonic missiles generally fall into two main categories. One type is the hypersonic glide vehicle, which is launched by a rocket and then glides through the atmosphere at very high speed toward its target. The other type is the hypersonic cruise missile, which uses advanced engines to maintain high speeds during its flight. Both types are designed to travel faster than traditional missiles and to follow less predictable paths. This combination of speed and maneuverability is what makes them particularly challenging to counter.

The working of a hypersonic missile involves several stages. In the case of a glide vehicle, the missile is first launched into the upper atmosphere using a rocket. After reaching a certain altitude, the glide vehicle separates and descends toward its target at very high speed. During this phase, it can adjust its direction, making its path difficult to predict. In the case of a cruise missile, the weapon uses a special engine that allows it to travel at hypersonic speeds within the atmosphere. These engines are designed to operate efficiently at very high speeds and temperatures.

The construction of hypersonic missiles requires advanced materials and precise engineering. At hypersonic speeds, the missile's surface experiences extreme heat due to air friction. This heat can be intense enough to damage or destroy ordinary materials. To address this, engineers use heat-resistant materials and special designs to protect the missile. The guidance systems must also be highly accurate, as controlling a vehicle at such high speed is very challenging. Because of these complexities, the development of hypersonic missiles is limited to countries with strong technological capabilities.

One of the main advantages of hypersonic missiles is their speed. Traveling at several times the speed of sound means that they can reach their targets very quickly. This reduces the time available for detection and response, making them difficult to intercept. Another advantage is their maneuverability. Unlike traditional ballistic missiles, which follow a predictable arc, hypersonic missiles can change direction during flight. This makes it harder for defense systems to track and engage them.

Hypersonic missiles also have a high level of precision. Their advanced guidance systems allow them to target specific locations with accuracy. This can make them effective for striking important targets such as military installations or infrastructure. In addition, their ability to travel long distances at high speed makes them well-suited for rapid-response situations.

However, the disadvantages and risks associated with hypersonic missiles are significant. One major concern is their unpredictability.

Because they can change direction and move at very high speeds, it is difficult to anticipate their path. This increases the risk of miscalculation and makes it harder to ensure that they are used in a controlled manner. The speed and maneuverability that provide advantages also pose challenges for safety and control.

Another issue is the difficulty of defense. Existing missile defense systems are often designed to deal with slower and more predictable threats. Hypersonic missiles can bypass these systems due to their speed and ability to change course. This creates a sense of vulnerability and may heighten tensions between countries. The need to develop new defense technologies can also lead to an arms race, where countries compete to build more advanced systems.

The impact on human life can be serious. In the event of a strike, the high speed of the missile means that there may be little or no warning. This can increase the risk to people in the target area. The impact can damage buildings, infrastructure, and essential services. In addition to physical harm, the presence of such weapons can create fear and uncertainty, especially in regions where tensions are high.

Human settlements can be affected by hypersonic missiles. In urban areas, a strike can disrupt transportation, communication, and utilities. The damage to infrastructure can make it difficult for communities to function and recover. Even the threat of such weapons can influence how cities plan their security and emergency response systems.

The economic impact is also important. Developing hypersonic missile systems requires significant investment in research, testing, and production. At the same time, the need to defend against these weapons can lead to additional costs. In the event of conflict, the damage caused by their use can result in high expenses for rebuilding and recovery. These factors can affect both national and global economies.

There are also ethical and legal considerations. The use of advanced weapons that are difficult to detect and intercept raises questions about fairness and control in conflict. International laws aim to regulate the use of force and protect civilians, but the rapid development of new technologies can make it challenging to keep these rules up to date. Discussions are ongoing about how to manage the risks associated with hypersonic weapons.

Another concern is the spread of this technology. As more countries develop hypersonic systems, there is a risk that they may become more widely available. This could increase the likelihood of their use and complicate global security. Efforts to control the spread of advanced weapons are an important part of international cooperation.

It is also worth noting that the research behind hypersonic technology has applications beyond military use. The study of high-speed flight can contribute to advances in transportation and space exploration. For example, similar principles may be used to develop faster aircraft or improve access to space. This shows that the same technology can have both positive and negative uses.

In conclusion, hypersonic missiles are a new and powerful type of weapon that combines extreme speed with the ability to change course mid-flight. They represent a major step forward in military technology and have the potential to change how conflicts are managed. While they offer advantages such as rapid response and precision, they also pose serious risks due to their unpredictability and the difficulty of defending against them. Their impact on human life, infrastructure, and global stability highlights the need for careful consideration and responsible use.

32. Improvised Explosive Devices (IEDs)

Improvised explosive devices, commonly known as IEDs, are homemade explosive weapons created using available materials rather than standardized military equipment. They are designed to cause damage by explosion and are often used in conflicts where one side lacks access to advanced weapons. These devices can vary widely in size, design, and method of use. Some are small and hidden in everyday objects, while others are larger and placed in vehicles or along roads. Their main feature is that they are improvised, meaning they are not produced in formal factories but assembled in informal or hidden settings.

The history of IEDs can be traced back to earlier forms of irregular warfare, where fighters used simple explosives to target stronger opponents. However, their widespread use became more evident in the late twentieth and early twenty-first centuries, especially in conflicts in which organized military forces faced insurgent groups. These groups often relied on IEDs because they were relatively inexpensive and could be made from commonly available materials. Over time, the design and use of these devices became increasingly advanced, with various triggering methods and concealment techniques developed.

An IED typically consists of a few basic components. There is an explosive material, which creates the blast. There is a container or casing that holds the explosive and may enhance its effect. There is also a triggering mechanism, which determines when the device will detonate. This trigger can be based on pressure, movement, time, or remote signals. In many cases, additional materials are added to increase the damage caused by the explosion. The design of each device depends on the purpose and the resources available.

The creation of IEDs involves combining these components in a way that allows the device to function as intended. Because they are improvised, the methods used can vary widely, and the level of sophistication can range from simple to complex. This variability makes IEDs difficult to predict and detect. The materials used may include everyday items, making it harder to identify potential threats. Due to the dangers involved, detailed information about their construction is restricted and not openly shared.

One of the main advantages of IEDs, from the perspective of those who use them, is their accessibility. They can be made from materials that are easier to obtain than those used for standard military weapons. This allows groups with limited resources to create weapons that can still have a significant impact. Another advantage is their ability to be hidden. IEDs can be placed in roads, vehicles, buildings, or other locations where they are not easily noticed. This allows them to be used for surprise attacks.

IEDs are also flexible in their use. They can be designed for different purposes, such as targeting vehicles, personnel, or infrastructure. Their placement can be adjusted based on the situation, making them adaptable to different environments. This flexibility has made them a common tool in certain types of conflicts.

However, the disadvantages and dangers of IEDs are very serious. One of the most important issues is their unpredictability. Because they are improvised, their behavior can be uncertain. They may not always function as intended and can sometimes detonate unexpectedly. This poses a high risk not only to the intended target but also to others in the area.

Another major concern is the impact on civilians. IEDs are often placed in public areas such as roads or marketplaces, where many people may be present. This increases the risk of harm to individuals who are not involved in the conflict. The hidden nature of these devices means that people may not be aware of the danger until it is too late. This has led to many incidents where civilians have been injured or killed.

The impact on human life is severe. Explosions caused by IEDs can result in injuries such as wounds from fragments, burns, and shock from the blast. These injuries can be life-threatening and may require immediate medical attention. In addition to physical harm, the psychological effect can be significant. The fear of hidden explosives can create stress and anxiety among people living in affected areas.

Human settlements can be deeply affected by the presence of IEDs. Roads and transportation routes may become unsafe, making it difficult for people to travel or access essential services. Buildings and infrastructure may be damaged by explosions, disrupting daily life. In some cases, communities may be forced to change their routines or avoid certain areas due to safety concerns. This can affect education, healthcare, and economic activities.

The economic impact of IEDs is also considerable. Damage to infrastructure requires resources for repair and rebuilding. Disruptions to transportation can affect trade and the movement of goods. Businesses may suffer if areas become unsafe or inaccessible. The cost of security measures and efforts to detect and remove IEDs can also be high. These factors can place a burden on both local and national economies.

There are also environmental effects to consider. Explosions can create debris and damage the surrounding area. In some cases, hazardous materials may be released, affecting soil and water. While the environmental impact may be more localized compared to larger weapons, repeated use of IEDs can still have noticeable effects over time.

From an ethical and legal perspective, the use of IEDs is widely condemned. International laws and agreements aim to protect civilians and limit the use of weapons that cause unnecessary harm. The hidden and unpredictable nature of IEDs makes them particularly dangerous and difficult to regulate. Their use raises serious questions about responsibility and the protection of human life.

Another important issue is the challenge of detection and removal. Because IEDs can be hidden in many different ways, identifying them requires specialized equipment and trained personnel. Efforts to clear areas of explosives can be time-consuming and dangerous. Even after a conflict ends, unexploded devices may remain, posing a continued risk to people.

It is also worth noting that the materials and knowledge used to create IEDs can have legitimate uses in other contexts. For example, controlled explosives are used in construction and mining. However, when these materials are used in uncontrolled and harmful ways, they pose a threat to safety and stability. This emphasizes the necessity of responsible utilization and governance.

In conclusion, improvised explosive devices are homemade weapons that rely on available materials and simple designs to create powerful explosions. They have become a common feature in certain types of conflicts due to their accessibility and adaptability. However, their unpredictability, potential for harm, and impact on civilians and communities make them highly dangerous. Their effects on human life, infrastructure, and society highlight the need for strong efforts to prevent their use and to protect people from their consequences.

33. Suicide Bombs

Suicide bombs are explosive devices that are carried or worn by an individual who deliberately detonates them, often in proximity to a chosen target. These attacks are usually planned to occur in crowded or important locations, such as public spaces, transportation hubs, markets, or gatherings. The person carrying the device becomes part of the delivery system, which makes the attack difficult to prevent once it reaches the target area. Because of this method, suicide bombings are considered one of the most dangerous forms of violence, especially for civilians.

The history of suicide bombings is connected to the broader history of conflict and political violence. While self-sacrificial acts in warfare have existed in earlier times, the modern use of suicide bombs became more visible in the late twentieth century. During this period, some groups began using this method to attack stronger opponents. These groups saw the approach as a way to bypass security and achieve maximum impact with limited resources. Over time, such attacks have occurred in different parts of the world, often linked to political, ideological, or religious conflicts. The global response has strongly condemned these acts due to their impact on innocent people.

A suicide bomb typically consists of an explosive material, a triggering mechanism, and a method of concealment. The explosive may be placed in a vest, belt, bag, or vehicle. The triggering mechanism can be manual, such as a switch, or based on other simple activation methods. The device is designed to detonate at a chosen moment, usually when the person is close to the intended target. Because these devices are often improvised, their design can vary widely depending on the available materials and the attackers' specific plan.

The creation of such devices involves combining explosive materials with a trigger and a carrying or concealment system. The design may be simple or complex, but the goal is to ensure the device can be transported and activated at the right time. Due to the serious risks and ethical concerns, detailed information about the construction of these devices is restricted and not openly shared.

One of the reasons suicide bombs are used is their ability to bypass many traditional security measures. Since a person carries the device, it can be moved through areas where other types of weapons might be detected or restricted. This makes it difficult for security systems to identify and stop the threat in advance. Another factor is the element of surprise. The attacker can choose the time and place of the explosion, which increases the potential impact.

However, the disadvantages and dangers of suicide bombings are extremely serious and far-reaching. The most important concern is the harm caused to civilians. These attacks often target places where people gather, meaning many of those affected are not involved in any conflict. The explosion can cause injuries such as wounds from fragments, burns, and damage from the blast itself. The loss of life and suffering caused by such attacks have led to widespread condemnation.

Another major issue is the psychological impact. The fear of sudden and unpredictable attacks can affect entire communities. People may feel unsafe in public spaces, which can change daily behavior and reduce social interaction. This fear can spread beyond the immediate area of an attack and influence how people think about safety and security.

The impact on human life is profound. Victims of such attacks may suffer long-term physical and emotional effects. The loss of loved ones significantly influences families and communities. The trauma associated with these events can last for many years and may require ongoing support and care.

Human settlements can also be significantly affected. Attacks in public areas can damage buildings, vehicles, and infrastructure. This can disrupt services such as transportation and communication. In some cases, areas may be closed or restricted after an attack, affecting daily life and economic activity. The need for increased security measures can also change how people move and interact in their communities.

The economic impact of suicide bombings can be considerable. Businesses in affected areas may suffer due to damage or reduced activity. Tourism can decline if a location is seen as unsafe. Governments may need to invest in security, emergency response, and recovery efforts. These costs can add up over time and affect overall economic stability.

There are also ethical and legal considerations. Suicide bombings are widely regarded as acts of terrorism and are condemned under international law. They violate principles that aim to protect civilians and limit harm in conflict. The intentional targeting of non-combatants is considered unacceptable by most societies and legal systems.

Another important issue is the challenge of prevention. Because these attacks involve individuals who are willing to sacrifice their own lives, traditional deterrence methods may be less effective. Preventing such attacks requires a combination of intelligence, security measures, and efforts to address underlying causes. Community awareness and cooperation can also play a role in identifying potential threats.

It is also important to consider the broader impact on society. Repeated attacks can lead to increased tension and mistrust between groups. This can affect social harmony and create divisions within communities. Efforts to respond to such threats must balance the need for security with the protection of individual rights and freedoms.

While the materials used in these devices may have legitimate uses in other contexts, their use in this way highlights the importance of responsible handling and regulation. Ensuring that materials and knowledge are not misused is an ongoing challenge that requires cooperation at local, national, and international levels.

In conclusion, suicide bombs are a form of explosive weapon carried by individuals and used in attacks that often target civilians. Their ability to bypass security and create sudden, high-impact events makes them particularly dangerous. The harm they cause extends beyond physical damage to include psychological, social, and economic effects. Understanding the nature of these attacks and their consequences is important for developing effective responses and promoting safety.

34. Booby Traps

Booby traps are concealed mechanisms intended to inflict injury when an individual interferes with or engages with a particular object or location. They are often placed in locations where people are likely to walk, touch, or move items, such as doorways, paths, buildings, or everyday objects. The key feature of a booby trap is that it is concealed and activated by an unexpected action, which makes it difficult to detect and avoid. These devices can be simple or complex, but they all rely on surprise to function.

The use of booby traps has a long history in warfare and conflict. In earlier times, people used simple traps, such as concealed pits or mechanical devices, to protect their territory or ambush enemies. As technology advanced, these traps became more sophisticated, especially during the twentieth century. During conflicts such as the Second World War and later wars, booby traps were used more widely. They were often placed in abandoned buildings, along roads, or in areas where opposing forces were expected to move. Over time, their use expanded in irregular warfare, where smaller groups used them to challenge larger, better-equipped forces.

A booby trap typically consists of three main elements: a triggering mechanism, an explosive or harmful component, and a method of concealment. The trigger is designed to activate when a person performs a specific action, such as stepping on a surface, opening a door, or picking up an object. The harmful component can vary, but in many cases, it involves an explosive that causes injury through blast or fragments. The concealment is what makes the trap effective, as it blends with the surroundings and does not attract attention.

The creation of booby traps involves combining these elements in a way that ensures the device will activate under the intended conditions. Because they are often improvised, the design can vary widely depending on the materials available and the environment in which they are used. This variation makes them unpredictable and difficult to identify. Due to the risks involved, detailed information about how such devices are made is restricted and not openly shared.

One of the main reasons booby traps are used is their ability to create surprise. Because they are hidden, they can affect people unaware of their presence. This can slow down movement, create fear, and make it harder for individuals or groups to operate in an area. Another factor is their relatively low cost and simplicity compared to more advanced weapons. This allows them to be used in situations where resources are limited.

Booby traps can also remain active for long periods if they are not discovered. This means that they can continue to pose a threat even after the conflict in which they were placed has ended. This lasting presence can make areas unsafe and require careful inspection and clearing before they can be used again.

However, the disadvantages and dangers of booby traps are very serious. One of the most important concerns is their lack of control. Once a trap is placed, it cannot distinguish between different people. This means that anyone who triggers it can be harmed, including civilians, aid workers, or even those who placed it. This inability to control who is affected makes booby traps particularly dangerous.

Another major issue is their hidden nature. Because they are designed to be difficult to detect, people may not realize they are in danger until it is too late. This increases the risk of accidental injury or death. Even trained personnel may find it challenging to identify and safely remove these devices, especially in areas where many traps have been placed.

The impact on human life can be severe. Booby traps can cause injuries such as wounds from fragments, burns, or blast damage. These injuries may require immediate medical attention and can have long-term effects. In addition to physical harm, the presence of hidden traps can create fear and stress, affecting the mental well-being of those who live or work in affected areas.

Human settlements can be significantly affected by the presence of booby traps. Areas where traps have been placed may become unsafe for daily activities. People may avoid certain roads, buildings, or fields, disrupting normal life. In some cases, entire communities may need to be relocated until the area is cleared. This can affect access to food, water, education, and healthcare.

The economic impact of booby traps is also important. Clearing areas of hidden devices requires time, specialized equipment, and trained personnel. This process can be costly and may delay the recovery of affected regions. Agriculture, construction, and other activities may be limited until the area is declared safe. This can reduce productivity and affect livelihoods.

There are also environmental effects to consider. Explosions can damage land and vegetation, and the presence of unexploded devices can prevent the use of natural resources. In some cases, animals may also be affected, either by triggering traps or by avoiding unsafe areas. The long-term presence of such hazards can change how land is used and managed.

From an ethical and legal perspective, the use of booby traps is highly controversial. International law aims to protect civilians and limit the use of weapons that cannot be effectively controlled. Because booby traps can harm anyone who encounters them, their use is often restricted or condemned. Efforts have been made to reduce their use and to promote the removal of existing hazards.

Another important issue is the challenge of detection and removal. Clearing booby traps is a dangerous task that requires careful planning and specialized skills. Teams that perform this work must use protective equipment and follow strict procedures to avoid accidents. Even with these measures, the risk remains high. This highlights the importance of preventing the use of such devices in the first place.

It is also worth noting that the concept of traps and triggering mechanisms has applications in other fields, such as safety systems and industrial processes. However, when used in harmful ways, these ideas pose a threat to human life and security. This shows the importance of using knowledge responsibly.

In conclusion, booby traps are hidden devices that cause harm when triggered by contact or movement. Their use in warfare and conflict has created serious risks for both military personnel and civilians. While they may offer certain tactical advantages, their unpredictability and long-lasting effects make them highly dangerous. Their impact on human life, communities, and the environment underscores the need for concerted efforts to prevent their use and clear affected areas. Understanding the nature of booby traps and their consequences is important for promoting safety and reducing harm.

35. Anti-Personnel Mines

Anti-personnel mines are explosive devices created to injure or take the lives of people. They are usually placed on or just below the ground in areas where movement is expected, such as paths, fields, or near buildings. These devices are triggered by a person's presence, pressure, or movement. Once activated, they explode, releasing force or fragments that can cause serious injury. Unlike many other weapons, anti-personnel mines do not require direct action at the moment of use. They remain hidden and active, waiting for someone to come into contact with them.

The history of anti-personnel mines goes back several centuries, but their widespread use began in the twentieth century, especially during major wars. During the Second World War, mines were used to protect positions, slow down enemy movement, and control access to certain areas. In later conflicts, their use increased, particularly in regions where fighting was prolonged. Over time, mines became more advanced, with different types designed for specific purposes. Some were made to explode under pressure, while others were designed to react to movement or disturbance. Their ability to remain active for long periods made them a lasting presence even after conflicts ended.

An anti-personnel mine typically includes an explosive charge, a casing, and a triggering mechanism. The casing protects the internal components and may also help control how the explosion spreads. The trigger can be activated by pressure, such as when someone steps on the mine, or by movement, such as a tripwire being disturbed. When triggered, the explosive charge detonates, producing a blast and often sending fragments outward. These fragments can travel in different directions, increasing the area of impact.

The creation of these mines involves careful design to ensure they function as intended. The materials used must be durable enough to remain effective over time, even in harsh environmental conditions.

The triggering mechanism must be sensitive enough to respond to human contact but stable enough to avoid accidental activation during placement. Because of the risks involved, the production and handling of such devices are controlled by governments and military organizations. Detailed information about their construction is not openly shared.

One of the main reasons anti-personnel mines have been used is their ability to control movement. They can be placed in strategic locations to prevent access or to guide movement in certain directions. This can be useful in defensive situations, where forces want to protect an area or slow down an approaching group. Mines can also be deployed over large areas, creating zones that are difficult to cross without risk.

Another advantage, from a military perspective, is their persistence. Once placed, they can remain active for a long time without further action. This means that they can continue to serve their purpose even after the area is no longer actively monitored. This characteristic has made them a tool for long-term area denial.

However, the disadvantages and dangers of anti-personnel mines are very serious and widely recognized. One of the most important concerns is their long-term impact. Mines can remain active for many years, or even decades, after they are deployed. This means that they continue to pose a risk long after a conflict has ended. People who return to their homes or begin using the land again may unknowingly encounter these hidden devices.

Another major issue is their effect on civilians. Anti-personnel mines do not distinguish between different individuals. They can harm anyone who comes into contact with them, including children, farmers, and aid workers. Many of the injuries caused by mines are severe and can result in permanent disability. This has led to widespread concern and efforts to reduce their use.

The impact on human life is significant. Injuries caused by mines often involve damage to limbs, which may require medical treatment and long-term care. In addition to physical harm, there can be emotional and psychological effects. Survivors may face challenges in returning to normal life, and families and communities are affected by the loss or injury of their members.

Human settlements can be deeply affected by the presence of anti-personnel mines. Areas that are contaminated with mines may become unsafe for living, farming, or travel. This can lead to displacement, as people are forced to leave their homes and seek safety elsewhere. Even when people remain, their daily activities may be limited by the risk of hidden explosives. This can affect access to food, water, education, and healthcare.

The economic impact is also important. Land that contains mines cannot be used safely for agriculture or development until it is cleared. This can reduce productivity and limit opportunities for growth. The process of clearing mines is complex, time-consuming, and costly. It requires trained personnel, specialized equipment, and careful planning. These efforts can take many years, especially in areas where large numbers of mines have been placed.

The environmental impact of anti-personnel mines is another concern. While the mines themselves may not always cause large-scale environmental damage, their presence can prevent land use and disrupt ecosystems. Animals may also be affected if they trigger mines. In some cases, natural habitats may remain untouched because they are unsafe for human activity, a situation that can have both positive and negative effects.

From an ethical and legal perspective, the use of anti-personnel mines has been widely debated. Many countries and organizations have raised concerns about their impact on civilians and their long-term effects. This has led to international agreements aimed at banning or restricting their use. These agreements encourage countries to stop using such weapons, destroy existing stockpiles, and assist in clearing affected areas.

Another important issue is the challenge of detection and removal. Mines are often difficult to locate because they are hidden and may be made of materials that are hard to detect. Clearing them requires careful work and carries risks for those involved. Despite advances in technology, complete removal can take a long time, and the danger may remain until the process is finished.

It is also worth noting that the technology used in mines, such as triggering mechanisms and explosive control, has applications in other fields when used safely. For example, controlled explosives are used in construction and mining for beneficial purposes. This highlights the importance of using such knowledge responsibly.

In conclusion, anti-personnel mines are hidden explosive devices designed to harm people and control movement in conflict zones. While they may offer certain advantages in military situations, their long-term risks and impact on civilians make them highly dangerous. Their presence can affect human life, communities, and economic development for many years. Understanding their effects and the challenges they create is important for promoting efforts to reduce their use and to support the clearing of affected areas.

36. Chemical Toxins (Ricin)

Chemical toxins such as ricin are highly poisonous substances that can cause serious harm to the human body even in very small amounts. Ricin is a natural toxin derived from the seeds of the castor plant, which is widely distributed worldwide. While the plant itself has useful applications, especially in the production of castor oil, ricin is a dangerous byproduct that can be extracted from the leftover material. Due to its remarkable potency, ricin has garnered significant attention in conversations regarding chemical hazards and security issues.

The history of ricin as a harmful substance is connected to both natural science and human conflict. Scientists first studied ricin in the late nineteenth century while researching plant-based toxins. Over time, its toxic nature became well understood. In the twentieth century, there were instances where ricin was used in small-scale attacks or assassination attempts. These cases were rare but drew attention because the substance could cause harm without being easily detected. Due to these concerns, ricin is now closely monitored under international agreements that regulate toxic substances.

Ricin works by interfering with the normal function of cells in the body. When it enters the body, it can prevent cells from producing the proteins they need to survive. This leads to cell damage and, in severe cases, organ failure. The effects of ricin depend on how it enters the body. It can be inhaled as a fine particle, ingested through contaminated food or water, or introduced through contact with broken skin. Each method of exposure can lead to different symptoms, but all forms are dangerous.

The process of obtaining ricin involves isolating it from castor bean material in industrial or laboratory settings. While castor oil production is a legitimate and common activity, the toxin is found in the leftover material after oil extraction. Because of the risks associated with ricin, its handling requires strict safety measures and is regulated by authorities. Detailed information about extraction and preparation is not publicly shared, as misuse could cause harm.

One of the reasons ricin is considered dangerous is its potency. Even a very small amount can cause serious health effects. This makes it a concern in situations where harmful substances might be used intentionally. Another factor is that ricin has no strong odor or taste, making it difficult to detect without proper equipment. These characteristics have contributed to its reputation as a substance of concern in security discussions.

However, there are no true advantages to the use of ricin as a weapon when viewed from a humanitarian or ethical perspective. While its potency might be seen as effective in causing harm, the suffering it creates and the risks it poses to both intended and unintended targets make it unacceptable. The lack of control over how it spreads or who it affects adds to its dangers.

The disadvantages and risks of ricin are severe. One of the main concerns is the difficulty of detecting exposure early. Symptoms may not appear immediately, which can delay treatment. Once symptoms develop, they can include difficulty breathing, severe discomfort, and damage to internal organs. Medical treatment focuses on supportive care, as there is no widely available specific antidote for ricin poisoning. Another major issue is the potential for accidental exposure. Because ricin can be present in materials associated with castor plant processing, improper handling can cause unintended harm. This highlights the importance of strict safety procedures in industries that deal with such materials.

The impact on human life can be serious and, in some cases, fatal. Individuals exposed to ricin may require immediate medical attention, and recovery can depend on the amount and method of exposure. In addition to physical effects, the fear associated with toxic substances can create anxiety and concern within communities.

Human settlements can be affected if a toxic substance is released in a populated area. Even a small-scale incident can lead to disruption, as authorities may need to secure the area, provide medical assistance, and investigate the situation. This can affect daily life and create a sense of uncertainty among residents. In some cases, areas may need to be cleaned or monitored to ensure safety.

The economic impact of incidents involving ricin can also be significant. Medical care, emergency response, and investigation efforts require resources. If businesses or public spaces are affected, there may be temporary closures or restrictions that impact local economies. In addition, the need for increased security and monitoring can lead to ongoing costs.

From an environmental perspective, ricin does not usually cause long-term, large-scale damage like some other hazardous substances. However, it can still pose risks in localized areas, especially if it contaminates surfaces or materials. Proper cleanup and disposal are necessary to prevent further exposure.

There are also ethical and legal considerations related to ricin. The use of toxic substances to harm others is widely condemned and is prohibited under international law. Agreements exist to regulate the production, storage, and handling of such materials. These rules are designed to protect people and prevent misuse. The strict control of ricin reflects the global effort to reduce the risk of chemical threats.

Another important point is the role of awareness and education. Understanding the dangers of toxic substances can help prevent accidents and misuse. Training for those who work with potentially hazardous materials is essential to ensure safety. Public awareness can also help people recognize risks and respond appropriately in case of an incident.

It is also worth noting that the castor plant, from which ricin is derived, has many beneficial uses. Castor oil is widely used in medicine, industry, and other fields. This shows that the same natural source can provide both useful products and dangerous substances. The key difference lies in how these materials are handled and used.

In conclusion, chemical toxins such as ricin are highly dangerous substances that can cause serious harm even in very small amounts. Derived from a natural source, ricin has been studied for many years and is known for its strong toxic effects. Although it has been involved in a few harmful incidents, its use is widely condemned for the suffering it causes and the risks it poses. Its impact on human life, communities, and security highlights the importance of strict control, responsible handling, and continued efforts to prevent misuse. Understanding the nature of such toxins and their effects is essential for promoting safety and protecting human well-being.

37. Weaponized Viruses

Weaponized viruses are viruses that have been studied, altered, or selected in ways that increase their ability to cause harm if they are deliberately released. In nature, viruses already exist and can cause illness. Still, when they are weaponized, the concern is that scientific knowledge could be misused to make them more dangerous or easier to spread. These viruses can infect living cells, multiply within the body, and, in some cases, spread from person to person. Because of this ability to reproduce and move through populations, they are often described as one of the most serious biological threats.

The idea of using disease as a weapon has existed for a long time. In ancient and medieval periods, there were reports of armies attempting to spread illness by contaminating water sources or using infected materials. However, the modern concept of weaponized viruses developed alongside advances in microbiology during the nineteenth and twentieth centuries. As scientists learned more about bacteria and viruses, they also understood how diseases spread and how they could be controlled. This knowledge brought benefits in medicine, but it also created concerns about misuse. During the twentieth century, some countries explored biological programs that studied pathogens for military purposes. Over time, strong international agreements were created to ban the development and use of biological weapons, including viruses.

A virus works by entering a host, such as a human or animal, and taking control of cells to produce more copies of itself. This process can damage tissues and lead to disease. In a weaponized context, the concern is not about creating viruses from nothing, but about modifying or selecting them to make them more harmful. This could include changes that affect how easily the virus spreads, how severe the illness becomes, or how long it survives in the environment. Such activities are highly restricted and regulated, and detailed information about them is not publicly shared.

The handling of viruses in laboratories requires strict safety measures. Scientists who work with infectious agents follow clear procedures to prevent accidental release and protect themselves and others. Specialized facilities are used to contain these organisms, with systems designed to control airflow, waste, and access. These precautions are essential because even natural viruses can pose risks if not handled properly. The idea of intentionally misusing such materials raises serious concerns, which is why international laws and agreements aim to prevent it.

One reason weaponized viruses are considered dangerous is their ability to spread. Unlike many other weapons with a limited area of effect, a virus can spread from one person to another, sometimes over long distances. This means that an initial release could lead to a wider outbreak. Another factor is the difficulty of detection. Early symptoms of viral infections can resemble those of common illnesses, delaying recognition and response. This delay can allow the virus to spread further before control measures are put in place.

However, there are no humanitarian advantages to using weaponized viruses. While their ability to spread might be seen as effective in causing disruption, the lack of control over their effects makes them extremely risky. Once a virus is released, it may not remain limited to the intended target and can affect many others, including those who were not meant to be involved. This uncontrollable nature makes them among the most concerning threats.

The disadvantages and risks are severe. One of the main issues is the potential for widespread illness. A virus that spreads easily can lead to large numbers of people becoming sick in a short time. This can overwhelm healthcare systems, making it difficult to provide treatment and care. Hospitals may become crowded, and resources such as medical supplies and personnel may be stretched.

Another major concern is the uncertainty of how a virus will behave. Even with scientific knowledge, it can be difficult to predict exactly how a virus will spread or how people will respond to it. Changes in the environment, population, and other factors can influence the outcome. This unpredictability increases the risk of unintended consequences.

The impact on human life can be profound. Viral outbreaks can cause illness, long-term health effects, and loss of life. In addition to physical health, there can be emotional and psychological effects. Fear of infection can change how people interact, work, and live. Communities may experience stress and uncertainty during outbreaks, which can affect overall well-being.

Human settlements can be significantly affected by the spread of a virus. In areas where outbreaks occur, movement may be restricted to reduce transmission. Schools, businesses, and public spaces may close temporarily. This can disrupt daily life and limit access to services. In some cases, people may need to isolate or quarantine, which can affect social and economic activities.

The economic impact of viral outbreaks can be large. Reduced activity in businesses, travel, and trade can lead to financial losses. Healthcare costs may increase as more people require treatment. Governments may need to invest in response measures, including testing, treatment, and public health campaigns. These factors can affect both local and global economies.

The environmental impact of viruses is generally less direct than that of physical weapons, but it can still be important. Changes in human activity during outbreaks can affect the environment, both positively and negatively. For example, reduced travel may lower pollution, while increased use of medical supplies may create waste. The overall effect depends on the scale and duration of the outbreak.

From an ethical and legal perspective, the use of weaponized viruses is strongly condemned. International agreements, such as those related to biological weapons, prohibit the development, production, and use of such agents. These agreements reflect a global understanding that biological threats pose serious risks to humanity. Efforts are ongoing to strengthen these rules and ensure compliance.

Another important issue is the risk of accidental release. Even in controlled environments, there is always a need for strict safety measures to prevent unintended exposure. Training, monitoring, and proper facilities are essential to reduce this risk. The possibility of accidents highlights the importance of responsible research and oversight.

It is also important to recognize that the study of viruses has many positive applications. Research in virology has led to vaccines, treatments, and a better understanding of diseases. This knowledge has saved many lives and improved public health. The challenge is to ensure that this knowledge is used for beneficial purposes and not for harm.

In conclusion, weaponized viruses pose a serious concern due to their ability to spread, cause illness, and disrupt society. While they are based on natural organisms, their potential misuse raises important questions about safety, ethics, and global security. Their impact on human life, communities, and economies can be far-reaching, especially if an outbreak is not controlled quickly. Understanding these risks and supporting efforts to prevent misuse are essential for protecting humanity and maintaining stability.

38. Weaponized Bacteria

Weaponized bacteria are harmful bacteria that have been studied, selected, or altered in ways that increase their ability to cause disease if they are used deliberately. In nature, many types of bacteria exist, and some can cause infections in humans, animals, or plants. When people talk about weaponized bacteria, they are referring to the misuse of scientific knowledge to make these organisms more dangerous or easier to spread. Because bacteria can grow and multiply inside the body and sometimes move from one person to another, they are considered a serious biological threat.

The idea of using disease as a weapon has long been known. In earlier periods of history, there were attempts to spread illness by contaminating water or using infected materials. However, the modern understanding of bacteria began in the nineteenth century, when scientists discovered how these microorganisms cause disease. This led to major improvements in medicine, such as the development of antibiotics and hygiene practices. At the same time, it raised concerns about how this knowledge could be misused. During the twentieth century, some countries explored biological programs that included research on bacteria for military purposes. These activities later led to international agreements that banned the development and use of biological weapons.

Bacteria are unicellular organisms capable of thriving in various environments. Some are harmless or even helpful, but others can cause disease by invading the body and producing harmful substances. In a weaponized context, the concern is not about creating bacteria from nothing, but about modifying or selecting more harmful strains. This could involve changes that increase their ability to spread, resist treatment, or survive in different conditions. Such activities are highly restricted, and detailed information about them is not publicly available.

The handling of bacteria in laboratories requires strict safety measures. Scientists work in controlled environments with limited access, and procedures are carefully followed to prevent exposure. Equipment such as protective clothing, sealed containers, and air control systems is used to reduce risk. These precautions are important because even natural bacteria can be dangerous if not handled properly. The possibility of misuse has led to strong regulations and oversight at both national and international levels.

One of the main reasons weaponized bacteria are considered dangerous is their ability to spread. Some bacterial infections can move from person to person through the air, water, or direct contact. This means that a small initial release could lead to a larger outbreak. Another factor is that bacteria can multiply quickly, increasing the number of infected individuals over time. In some cases, certain bacteria can survive outside the body for a period, increasing the risk of transmission.

Another concern is resistance to treatment. Some bacteria have developed resistance to antibiotics, which are the medicines used to treat bacterial infections. If a harmful strain is resistant, it becomes more difficult to control and treat the disease. This increases the risk of serious health outcomes and makes outbreaks harder to manage. The possibility of resistance is one reason the misuse of bacteria is considered particularly dangerous.

There are no real advantages to using weaponized bacteria from a humanitarian or ethical perspective. While their ability to spread might be seen as effective in causing disruption, the lack of control over their effects makes them extremely risky. Once bacteria are released, they may affect people beyond the intended target, including those who had no involvement in any conflict. This uncontrollable nature makes them among the most concerning threats.

The disadvantages and risks are very serious. One of the main issues is the potential for widespread illness. A bacterial outbreak can lead to many people becoming ill, placing a heavy burden on healthcare systems. Hospitals may become overcrowded, and there may be shortages of medical supplies and staff. This can make it difficult to provide proper care for everyone who needs it.

Another major concern is the unpredictability of how an outbreak will develop. Factors such as population density, travel, and environmental conditions can influence how bacteria spread. Even with planning and knowledge, it can be difficult to predict the exact course of an outbreak. This uncertainty can make it harder to respond effectively and can increase the overall impact.

The effect on human life can be severe. Bacterial infections can cause a range of symptoms, from mild illness to serious, potentially fatal conditions. Some infections can have long-term effects on health, even after recovery. In addition to physical harm, there can be emotional and psychological effects. Fear of infection can affect how people interact with others and can lead to stress and anxiety.

Human settlements can be significantly affected by the spread of harmful bacteria. In areas where outbreaks occur, measures may be taken to limit movement and reduce contact between people. Schools, workplaces, and public spaces may be closed temporarily. This can disrupt daily life and make it harder for people to access services such as education and healthcare. In some cases, entire communities may need to adjust their routines to reduce risk.

The economic impact of bacterial outbreaks can be substantial. Reduced activity in businesses, travel, and trade can lead to financial losses. Healthcare costs may increase as more people require treatment. Governments may need to invest in public health measures, including testing, treatment, and education campaigns. These costs can affect both local and global economies.

From an environmental perspective, bacteria can affect ecosystems if they spread among animals or contaminate water sources. While the direct environmental impact may be less visible than that of physical weapons, it can still be important. Changes in human activity during outbreaks can also influence the environment, for example, through increased waste or reduced industrial activity.

There are also ethical and legal considerations. The use of bacteria as a weapon is prohibited under international agreements that aim to protect human life and prevent the misuse of biological agents. These agreements reflect a shared understanding that biological threats pose serious risks to global health and security. Efforts are ongoing to strengthen these rules and ensure that they are followed.

Another important issue is the risk of accidental release. Even in controlled settings, there is always a need for strict safety procedures to prevent unintended exposure. Training, monitoring, and proper facilities are essential to reduce this risk. The possibility of accidents highlights the importance of responsible research and strong oversight.

It is also important to recognize that the study of bacteria has brought many benefits to humanity. Advances in microbiology have led to better treatments, vaccines, and public health practices. Many bacteria are also used in industry, agriculture, and environmental management in positive ways. This shows that the same knowledge can be used for both helpful and harmful purposes, depending on how it is applied.

In conclusion, weaponized bacteria are harmful microorganisms that could be misused to cause disease and disruption. Their ability to spread, multiply, and sometimes resist treatment makes them a serious concern. While they are based on natural organisms, their potential misuse raises important questions about safety, ethics, and global cooperation. Their impact on human life, communities, and economies can be wide-ranging, especially if an outbreak is not controlled quickly. Understanding these risks and supporting efforts to prevent misuse are essential for protecting public health and maintaining stability.

39. Anti-Satellite Weapons

Anti-satellite weapons, often called ASAT weapons, are systems designed to damage, disable, or destroy satellites orbiting Earth. Satellites play a very important role in modern life. They support communication, navigation, weather forecasting, scientific research, and many other services. Because of this, any weapon that targets satellites can have effects that extend far beyond the immediate target. Anti-satellite weapons are considered part of space-related military technology, and their development reflects the growing importance of space in global systems.

The history of anti-satellite weapons began during the Cold War, when major powers began exploring how space systems could be used in military operations. As satellites became more common, countries began to consider ways to protect their own systems and, at the same time, to disrupt others'. Early experiments involved missiles launched from the ground to intercept satellites in orbit. Over time, various methods were developed, including systems capable of interfering with satellite signals or damaging their electronics. In recent years, several countries have demonstrated anti-satellite capabilities, raising concerns about the security of space-based infrastructure.

Anti-satellite weapons can take different forms. One common type is a missile launched from the ground or another platform and directed toward a satellite. The missile travels at high speed and collides with the target, destroying or damaging it. Another type involves electronic interference, where signals are used to disrupt communication between a satellite and its control systems. There are also concepts involving directed energy, such as lasers, that can affect satellite sensors or components. Each type of system uses different technology, but all are designed to reduce or eliminate a satellite's function.

The development of anti-satellite systems requires advanced technology and precise coordination. The weapon must be able to track a satellite moving at very high speed in orbit and reach it accurately. This involves sensors, guidance systems, and powerful propulsion. Because of the complexity and risks, these systems are developed under strict control, and detailed technical information is not widely available. The challenge of operating in space adds another layer of difficulty, as factors such as distance, speed, and environmental conditions must all be considered.

One of the main advantages of anti-satellite weapons, from a military perspective, is their ability to disrupt critical systems. Satellites are used for communication, navigation, and surveillance. If these systems are disabled, it can affect a country's ability to coordinate its activities. This may present a competitive advantage in certain contexts. Another advantage is that these weapons can target systems that are otherwise difficult to reach, as satellites operate far above the Earth.

However, the disadvantages and risks of anti-satellite weapons are very serious. One of the most important concerns is the creation of space debris. When a satellite is destroyed, it can break into many pieces that continue to travel in orbit at high speed. These pieces can collide with other satellites, causing further damage. This chain reaction can increase the amount of space debris, making it more dangerous for all satellites. This problem is often referred to as a long-term risk because debris can remain in orbit for many years.

Another major issue is the global impact. Satellites are used by many countries and support services that people rely on every day. For example, navigation systems help with transportation, communication satellites support phone and internet services, and weather satellites provide important information for safety and planning. If these systems are disrupted, the effects can be felt worldwide, not just in one area. This makes anti-satellite weapons different from many other weapons, as their impact is not limited to a specific location.

The impact on human life can be indirect but still significant. If satellite systems are affected, people may experience disruptions in communication, navigation, and access to information. This can affect emergency services, transportation, and daily activities. In some cases, it can also affect safety, for example, if navigation systems are unavailable on aircraft or ships.

Human settlements can also be affected by disruptions to satellite systems. Cities depend on communication networks and data systems that rely on satellites. If these systems are interrupted, it can affect services such as banking, transportation, and utilities. This can create challenges for both individuals and organizations, especially in highly connected environments.

The economic impact of anti-satellite weapons can be very large. Satellites represent a major investment in technology and infrastructure. Destroying or disabling them can lead to financial losses and require replacement systems, which can be expensive and time-consuming. In addition, service disruptions can affect businesses, trade, and industries that depend on reliable communication and data.

From an environmental perspective, space debris is one of the most serious concerns. Unlike debris on Earth, which can often be cleaned up, space debris is difficult to remove. It can remain in orbit for a long time and continue to pose a risk to other objects. This can affect not only military systems but also scientific and commercial satellites. The long-term presence of debris can limit the safe use of space for future activities.

There are also ethical and legal considerations. Space has often been viewed as a shared environment that should be used for peaceful purposes. The development and testing of anti-satellite weapons raise questions about how space should be managed and protected. International discussions and agreements aim to reduce the risks associated with such activities and promote cooperation. However, the rapid development of technology makes it challenging to keep rules up to date.

Another important issue is the risk of escalation. Actions in space can prompt responses from other countries, heightening tension and uncertainty. Because satellites are connected to many systems, any disruption can have wide effects. This makes it important to consider the broader consequences of using such weapons.

It is also worth noting that the technology used in anti-satellite systems has connections to other fields. Tracking, guidance, and communication technologies are used in many peaceful applications, including space exploration, weather monitoring, and global communication. This shows that the same knowledge can be used for both beneficial and harmful purposes.

In conclusion, anti-satellite weapons are systems designed to target satellites in orbit. They have developed over time as the importance of space in modern life has grown. While they may offer certain strategic advantages, their risks and consequences are significant. The creation of space debris, the disruption of global systems, and the potential for widespread impact make them a serious concern. Understanding these weapons and their effects is important for promoting responsible use of space and ensuring that it remains a safe and useful environment for all.

40. Space-Based Weapons

Space-based weapons are systems placed in orbit around Earth that can affect targets on the ground, in the air, or in space. These weapons are often discussed in theory and research, and most are not widely deployed in practice. The idea behind them is to use the positions of objects in space to gain advantages such as faster response times, wider coverage, and the ability to reach distant targets. Because satellites already orbit the Earth for communication and observation, the concept of placing weapons in space has been studied for many years.

The history of space-based weapons is closely linked to the development of space technology during the twentieth century. When countries began launching satellites, they also began to think about how space could be used for military purposes. During the Cold War, both major powers explored ideas related to space defense and offense. Some proposals included systems capable of intercepting missiles or striking targets from orbit. However, concerns about safety, cost, and international stability limited the development of such weapons. Agreements were also created to prevent the placement of certain types of weapons in space, especially those involving nuclear technology. As a result, most space-based weapon concepts have remained at the level of research and discussion rather than being deployed widely.

Space-based weapons can take different forms depending on the concept. Some ideas involve platforms in orbit that could release projectiles toward targets on Earth. Others focus on systems that could disable or interfere with satellites or other space objects. There are also concepts involving directed energy, such as beams that could affect targets at a distance. Each of these ideas relies on the unique position of space systems, which allows them to operate above the Earth and cover large areas.

The development of such systems requires advanced technology and careful design. Objects in orbit move at very high speeds and must be controlled with precision. Any system placed in space must be able to withstand extreme conditions, including temperature changes, radiation, and the vacuum of space. Guidance and communication systems must be reliable, as signals need to travel long distances. Because of these challenges, developing space-based weapons requires significant resources and expertise. Detailed technical information is not widely available, and research is usually conducted under strict control.

One of the main advantages of space-based weapons is their reach. From orbit, a system can potentially cover large portions of the Earth. This allows for rapid response to events in different locations. Another advantage is the speed at which an effect can be delivered. Because objects in space move quickly, a system could reach a target faster than some traditional methods. This can reduce the time available for detection and response.

Space-based systems can also provide a unique perspective. Being above the Earth allows for observation and tracking over wide areas. This can support coordination and planning. In theory, such systems could be used to intercept threats or provide support for defensive operations. These potential benefits are part of the reason why research in this area continues.

However, the disadvantages and risks of space-based weapons are very serious. One of the main concerns is the difficulty of defense. If a system in orbit is used to target a location on Earth, it may be hard to detect and intercept in time. This creates a sense of vulnerability, as traditional defense systems are not always designed to handle threats from space.

Another major issue is the risk to space itself. Placing weapons in orbit increases the possibility of conflict in space. If systems are damaged or destroyed, they can create debris that remains in orbit for a long time. This debris can pose a danger to other satellites and space missions. Over time, this could make certain areas of space more difficult to use safely.

The impact on human life can be indirect but significant. If space-based weapons are used, they could disrupt services that people depend on, such as communication, navigation, and weather forecasting. These services are supported by satellites, and any interference can affect daily life. In addition, the potential for rapid, hard-to-detect attacks can heighten fear and uncertainty.

Human settlements can also be affected by the consequences of space-based systems. If essential services are disrupted, cities may face challenges in managing transportation, communication, and utilities. Emergency services may be affected if communication systems are not functioning properly. The broader impact can extend beyond the immediate area of any direct action.

The economic impact of space-based weapons can be very large. Developing and maintaining systems in orbit requires significant investment in technology, launches, and operations. In addition, the risk of satellite damage can lead to financial losses. Many industries depend on space-based services, and disruptions can affect trade, business, and global markets. The cost of replacing damaged systems can also be high.

From an environmental perspective, space debris is one of the most important concerns. Unlike debris on Earth, which can often be managed, space debris can remain in orbit for many years. It can travel at high speeds and collide with other objects, creating more debris. This chain reaction can increase the risk to all space activities, including scientific research and exploration.

There are also ethical and legal considerations. Space has often been viewed as a shared environment that should be used for peaceful purposes. International agreements have been created to limit certain activities and promote cooperation. The idea of placing weapons in space raises questions about how to balance security with the need to protect this shared environment. Discussions continue about how to manage these challenges and prevent conflict.

Another important issue is the potential for escalation. The introduction of weapons into space could lead to increased competition and tension between countries. This could result in further development of similar systems, creating a cycle that is difficult to control. The global nature of space means that actions taken by one country can affect others, making cooperation essential.

It is also worth noting that the technology used in space-based systems has many positive applications. Satellites support communication, navigation, weather monitoring, and scientific research. Advances in space technology have improved life on Earth in many ways. This shows that the same knowledge can be used for beneficial purposes when applied responsibly.

In conclusion, space-based weapons are systems that would operate from orbit and could affect targets on Earth or in space. While they offer theoretical advantages such as wide reach and rapid response, they also present serious risks. The difficulty of defense, the creation of space debris, and the potential for global impact make them a major concern. Their development raises important questions about safety, ethics, and the future of space activities. Understanding these issues is essential for ensuring that space remains a safe and shared environment for all.

41. Kinetic Bombardment

Kinetic bombardment, sometimes described by the informal name "rods from the sky," is a concept for a weapon that relies on speed and mass rather than explosives to create damage. The basic idea is simple. A dense metal rod is released from a platform in orbit and allowed to fall toward the Earth at extremely high speed. As it travels, gravity pulls it downward while its velocity increases. When it reaches the ground, the energy created by its speed and weight produces a powerful impact. This impact can damage structures and penetrate deep into the ground without requiring any explosive material.

The idea of kinetic bombardment comes from the broader study of physics and space technology. Scientists have long understood that an object moving at high speed carries a large amount of kinetic energy. This energy increases with both mass and velocity. During the twentieth century, as space exploration developed, some researchers began to consider how objects in orbit could be used in new ways. The concept of using dense projectiles from space was discussed in military research during the Cold War, though it was never fully developed into an operational system. It has remained mostly theoretical due to technical, financial, and legal challenges.

A typical concept for this type of weapon involves placing long, dense metal rods in orbit. Materials such as tungsten are often mentioned in discussions because they are very dense and can withstand high temperatures. These rods would be stored on a satellite or another space-based platform. When needed, a rod would be released and guided toward a target on Earth. As it enters the atmosphere, it would experience extreme heat due to friction with the air. The design would need to ensure that the rod remains intact during this phase. By the time it reaches the surface, it would be traveling at a very high speed, delivering a strong impact.

The construction of such a system would require advanced engineering. The rods must be shaped to reduce air resistance and maintain stability during descent. The platform in orbit must be able to store, release, and guide the rods accurately. Guidance systems would be needed to ensure that the projectile reaches the intended target. Because of the extreme conditions involved, materials must be carefully selected to handle heat and stress. These challenges make the system complex and expensive to develop, which is one reason why it has not been widely implemented.

One of the main advantages of kinetic bombardment is that it does not rely on explosives. This reduces some of the risks associated with storing and handling explosive materials. The damage is caused purely by the projectile's kinetic energy, which can be very high at high speeds. Another advantage is precision. With proper guidance, the rod could be directed at a specific target, potentially reducing unintended damage to surrounding areas compared to some other weapons.

Speed is another important factor. Because the projectile falls from space, it can reach its target quickly, leaving little time for detection or response. This makes it difficult for traditional defense systems to intercept. The high velocity also increases the impact's effectiveness, allowing the rod to penetrate strong structures or underground facilities.

However, the disadvantages and challenges of kinetic bombardment are significant. One of the main issues is the cost. Launching materials into space requires a large amount of energy and resources. Maintaining an orbital platform capable of storing and deploying such projectiles adds to the expense. This makes the system less practical compared to other weapons that can be deployed from the ground or air.

Another concern is the technical difficulty. Ensuring that the rod survives re-entry into the atmosphere without breaking apart is a major challenge. The guidance system must also be highly accurate, as even small errors at high speed can cause it to miss the target. These technical requirements make the system complex and difficult to operate.

The impact on human life can be serious if such a weapon were used. The force of the impact can cause destruction comparable to a large explosion, even though no explosives are involved. People in the affected area could be injured or killed, and buildings and infrastructure could be damaged. The sudden nature of the impact means that there may be little warning.

Human settlements could also be affected in various ways. A high-speed impact can damage roads, buildings, and essential services such as water and electricity. This can disrupt daily life and require time and resources for recovery. In urban areas, the effects could be more severe due to the density of structures and population.

The environmental impact is another consideration. The impact of a dense object at high speed can disturb the ground and create debris. While it may not produce chemical pollution like some other weapons, the physical damage to land and structures can still be significant. The launch and maintenance of space-based systems also have environmental effects related to energy use and materials.

From an economic perspective, the cost of developing and maintaining kinetic bombardment systems is very high. In addition to the expense of launching equipment into space, there are ongoing costs for operation, monitoring, and maintenance. The damage from an impact can also lead to economic losses in affected areas, including the costs of rebuilding and restoring services.

There are also ethical and legal considerations. The idea of placing weapons in space raises questions about the use of a shared environment. International agreements have been created to limit certain activities in space and promote peaceful use. The deployment of such weapons could challenge these agreements and lead to increased tension between countries. The potential for rapid and hard-to-defend attacks also raises concerns about stability and security.

Another important issue is the risk of escalation. The introduction of space-based weapons could encourage other countries to develop similar systems or countermeasures. This could lead to competition and increase the risk of conflict. Because space systems are connected to many aspects of modern life, any disruption can have wide-reaching effects.

It is also worth noting that the scientific principles behind kinetic bombardment have positive applications. The study of high-speed motion and materials is used in fields such as aerospace engineering and space exploration. These applications contribute to advances in technology and knowledge. This shows that the same principles can be used in ways that benefit society.

In conclusion, kinetic bombardment is a concept that uses high-speed metal projectiles from space to create powerful impacts without explosives. It is based on the simple idea of converting speed and mass into energy, but its practical implementation involves complex technology and high costs. While it offers advantages such as precision and speed, it also presents serious challenges and risks. Its potential impact on human life, infrastructure, and global stability makes it a subject of concern. Understanding this concept helps highlight the importance of responsible use of technology and the need to consider the broader effects of advanced weapons systems.

42. Electronic Warfare Systems

Electronic warfare systems are technologies used to control, disrupt, or interfere with electronic signals such as communication, radar, and navigation. These systems do not rely on physical force, such as explosions or projectiles. Instead, they work by affecting the electronic environment in which modern systems operate. Since many military and civilian systems depend on electronic signals, electronic warfare can have a wide and powerful impact without causing visible damage.

The development of electronic warfare began in the early twentieth century, alongside the growth of radio communication. During the First World War, armies began using radio signals to send messages, and soon after, they also began intercepting or blocking them. By the Second World War, electronic warfare had become more advanced. Radar systems were used to detect aircraft, and efforts were made to confuse or jam them. Over time, technology improved, and electronic warfare became a key part of military operations. Today, it includes a wide range of tools and techniques that can affect communication, sensing, and control systems.

Electronic warfare systems generally fall into three main categories. The first is electronic support, which involves detecting and analyzing signals. This helps identify the presence and location of systems such as radar or communication devices. The second is an electronic attack, which aims to disrupt or turn off these systems. This can include jamming signals, sending false information, or interfering with the normal operation of equipment. The third is electronic protection, which focuses on defending systems against such interference and ensuring that they continue to function.

The operation of electronic warfare systems involves transmitters, receivers, and software. These systems can send out signals that interfere with or override existing signals. For example, a jammer can transmit noise on the same frequency as a communication system, making it difficult or impossible to understand messages. More advanced systems can mimic real signals to confuse equipment or provide false information. Because these technologies rely on knowledge of electronic systems, their development requires expertise in engineering, computing, and signal processing. Detailed technical information is usually restricted due to security concerns.

One of the main advantages of electronic warfare is that it can achieve its goals without physical destruction. Disrupting communication or radar can reduce the effectiveness of an opponent's systems. This can provide a strategic advantage while limiting visible damage to infrastructure. Another advantage is its range. Electronic signals can cover large areas, allowing a single system to affect multiple targets at once.

Electronic warfare systems are also flexible. They can be adjusted to target specific frequencies or types of signals, making them adaptable to different situations. In some cases, they can be used to gather information that helps with planning and decision-making. This combination of disruption and information gathering makes electronic warfare an important part of modern operations.

However, the disadvantages and risks of electronic warfare are also significant. One major concern is the impact on civilian systems. Many everyday services, such as mobile communication, internet access, and navigation, rely on electronic signals. If these signals are disrupted, it can affect people who are not involved in any conflict. For example, interference with navigation systems can affect transportation, including aircraft and ships.

Another issue is the difficulty of control. Electronic signals do not always stay within a defined area, and interference can spread beyond the intended target. This can lead to unintended effects on other systems. The invisible nature of electronic warfare also makes it harder to detect and understand what is happening, creating confusion.

The impact on human life can be indirect but important. While electronic warfare does not usually cause physical injury, it can affect systems that people depend on for safety and daily life. For example, disruption of communication can affect emergency services, making it harder to respond to accidents or crises. Interference with navigation can create risks in transportation.

Human settlements can be affected by disruptions to essential services. Cities rely on communication networks, power systems, and data systems that are connected through electronic signals. If these systems are affected, it can lead to problems such as loss of connectivity, service delays, and difficulty managing resources. In highly connected areas, the effects can spread quickly and affect many people.

The economic impact of electronic warfare can be significant. Disruptions to communication and data systems can affect businesses, trade, and financial transactions. Companies that rely on digital systems may experience losses if their operations are interrupted. In addition, the cost of developing, maintaining, and defending against electronic warfare systems can be high.

From an environmental perspective, electronic warfare does not usually cause direct physical damage to land or ecosystems. However, its indirect effects, such as disruptions to infrastructure, can influence how resources are used and managed. The production and operation of electronic systems also require energy and materials, which have their own environmental impact.

There are also ethical and legal considerations. The use of electronic warfare raises questions about the protection of civilian systems and the limits of interference. International laws aim to reduce harm and protect essential services, but the rapid development of technology can make it challenging to keep rules up to date. The invisible nature of electronic warfare adds to these challenges, as it may not always be clear who is responsible for an action.

Another important issue is the potential for escalation. Because electronic warfare can affect many systems at once, it can provoke strong reactions and heighten tensions between countries. The ability to disrupt critical infrastructure can be seen as a serious threat and may lead to further action. This highlights the need for careful management and communication.

It is also important to recognize that the technology used in electronic warfare has many positive uses. Communication systems, radar, and signal processing are essential for modern life. They support services such as broadcasting, navigation, and emergency response. Advances in these areas have improved safety, efficiency, and connectivity. This shows that the same technology can be used for both helpful and harmful purposes, depending on how it is applied.

In conclusion, electronic warfare systems are tools that use electronic signals to disrupt, control, or gather information about communication and radar systems. They have developed over many years and are now a key part of modern operations. While they offer advantages such as flexibility and reduced physical damage, they also pose risks due to their wide reach and indirect effects on civilian life. Their impact on communication, infrastructure, and global stability highlights the need for responsible use and careful regulation. Understanding these systems helps recognize both their capabilities and the challenges they pose to humanity and modern society.

43. Smart Bombs

Smart bombs are guided weapons designed to hit specific targets with a high level of accuracy. Unlike traditional bombs, which fall freely and depend on gravity and general aim, smart bombs use guidance systems to adjust their path during flight. This allows them to strike a chosen location more precisely. They are widely used in modern military operations, where accuracy is important for reducing unintended damage and increasing the chance of hitting the intended objective.

The development of smart bombs began during the twentieth century, especially during and after the Second World War. Early bombs were unguided, and their effectiveness depended on the pilot's skill and the conditions at the time of release. As technology improved, engineers began to explore ways to guide bombs toward their targets. During later conflicts, such as the Vietnam War, early versions of guided bombs were used. These systems were not perfect, but they showed that precision could be improved. Over time, advances in electronics, computing, and navigation led to more reliable and accurate smart bombs. Today, they are a standard part of many military forces.

A smart bomb typically includes a standard explosive body along with a guidance system and control surfaces. The guidance system may use various methods to locate and track the target. Some systems rely on signals from satellites to determine their position and adjust their path. Others use lasers, in which a target is marked with a laser beam, and the bomb follows the reflected signal. Some systems use cameras or sensors to recognize and track a target. The control surfaces, such as small fins, allow the bomb to change direction during flight based on the guidance system's instructions.

The process of creating a smart bomb involves combining these components into a single system. The explosive part is similar to that of a traditional bomb, but the addition of guidance and control systems makes the weapon more complex. Engineers must ensure that all parts work together correctly, even under conditions such as high speed and strong forces during release and flight. Because of the technical nature of these systems, specialized organizations handle their design and production, and detailed information is not publicly shared.

One of the main advantages of smart bombs is their accuracy. By guiding the bomb toward a specific target, they can reduce the likelihood of missing. This can make them more effective in achieving their intended purpose. Another advantage is the potential to reduce unintended damage to surrounding areas. Because the weapon is directed toward a precise location, it may limit the impact on nearby structures compared to less accurate weapons.

Smart bombs can also be used from a distance. Aircraft can release them without flying directly over the target, thereby reducing risk to personnel. The ability to adjust the bomb's path during flight provides flexibility, especially when conditions change.

However, the disadvantages and risks of smart bombs are important to consider. One issue is that accuracy does not guarantee safety. Even with precise targeting, unintended effects can still occur. For example, if the target is located in a populated area, nearby people and structures may still be affected. This means that while smart bombs can reduce collateral damage, they cannot eliminate it.

Another concern is the reliance on technology. Guidance systems depend on signals, sensors, and data. If these systems are disrupted or inaccurate, the bomb may not reach the intended target. This can reduce effectiveness and increase the risk of unintended impact. Environmental factors, such as weather or obstacles, can also affect performance.

The impact on human life can be serious. Smart bombs are designed to destroy specific targets, and their use can result in loss of life and injury. Even when used with precision, the presence of people near the target can lead to harm. The psychological impact of such attacks can also be significant, as the ability to strike accurately may create fear and uncertainty.

Human settlements can be affected by the use of smart bombs, especially when targets are located within or near communities. Damage to buildings, infrastructure, and services can disrupt daily life. In some cases, essential systems such as water, electricity, and transportation may be affected. Recovery from such damage can take time and resources.

The economic impact of smart bombs includes both the cost of development and the effects of their use. Developing advanced guidance systems requires investment in research and technology. At the same time, the damage caused by their use can lead to expenses for rebuilding and restoring affected areas. Businesses and local economies may also be affected if infrastructure is damaged or activity is disrupted.

From an environmental perspective, the impact of smart bombs is similar to that of other explosive weapons. The explosion can damage land, structures, and natural surroundings. Debris and pollution may result from the destruction of buildings or equipment. While the precision of smart bombs may limit the area affected, the environmental effects within that area can still be significant.

There are also ethical and legal considerations. The use of smart bombs is often discussed in terms of reducing harm by increasing accuracy. However, questions remain about how and where they are used. International laws aim to protect civilians and limit the impact of conflict on non-combatants. The use of precise weapons does not absolve one of the responsibility to follow these rules. Decisions about targeting and use are subject to careful consideration and oversight.

Another important point is the role of smart bombs in modern warfare. Their development reflects a broader trend toward precision and technology in military operations. This has changed how conflicts are conducted, placing greater emphasis on targeted actions. While this can reduce some types of damage, it also raises questions about the long-term effects and the balance between technology and human judgment.

It is also worth noting that the technology used in smart bombs has connections to other fields. Guidance systems, sensors, and navigation technologies are used in many civilian applications, such as aviation, transportation, and communication. These technologies have improved safety and efficiency in everyday life. This shows that the same scientific knowledge can be used for both beneficial and harmful purposes.

In conclusion, smart bombs are guided weapons that use advanced technology to improve accuracy and target specific locations. They represent a significant development in military systems, offering advantages such as precision and flexibility. However, they also carry risks, including unintended harm and reliance on complex technology. Their impact on human life, communities, and the environment highlights the need for careful and responsible use. Understanding smart bombs helps to provide a clearer view of how modern technology influences conflict and its effects on society.

44. Fuel-Air Explosives

Fuel-air explosives are weapons that create a powerful blast by using a cloud of fuel mixed with air. Instead of relying only on a solid explosive, they spread a fine mist or gas of fuel into the surrounding air and then ignite it. When the fuel cloud burns, it produces a large and intense explosion. This type of weapon is often grouped with thermobaric weapons because both use atmospheric oxygen to intensify the blast. The result is a strong pressure wave and heat that can affect a wide area.

The idea behind fuel-air explosives comes from basic science. When fuel mixes well with air and is ignited, it burns rapidly, releasing energy. This principle is seen in everyday situations, such as in engines and controlled industrial processes. However, in the case of fuel-air explosives, the reaction is uncontrolled and occurs very quickly, resulting in a sudden, powerful release of energy. This concept began to be explored more seriously in the mid-twentieth century, when advances in chemistry and engineering made it possible to design systems that could disperse fuel in a controlled way before ignition.

The historical development of these weapons occurred in the late twentieth century. Military research looked for ways to increase the effectiveness of explosions without simply increasing the size of traditional explosives. Fuel-air explosives offered a different approach. Instead of one concentrated blast from a compact charge, they yielded a wider and more lasting pressure effect. These weapons were tested and used in certain conflicts, particularly in situations where a large area needed to be affected. Over time, improvements were made in how the fuel cloud is dispersed and ignited.

A fuel-air explosive generally works in two stages. First, the weapon releases a cloud of fuel into the air. This fuel can be in the form of a fine liquid mist or gas. The goal is to mix it evenly with the surrounding air to create a flammable cloud. After a short delay, a second action ignites this cloud. When ignition occurs, the fuel burns very quickly, producing a large fireball and a strong pressure wave. This pressure wave can travel outward, damaging structures and objects in its path.

The construction of such a weapon involves a fuel container, a system to disperse it, and a mechanism to ignite it at the right time. The design must ensure that the fuel spreads properly and that ignition occurs when the mixture is most effective. Engineers must consider factors such as weather conditions, air movement, and the environment in which the weapon is used. Because of the complexity and risks, the development of these systems is controlled, and detailed technical information is not openly shared.

One of the main advantages of fuel-air explosives is their wide-area effect. The blast can cover a larger area than some traditional explosives of similar size. This makes them effective in situations where the target is spread out or located in open areas. Another advantage is the strong pressure wave, which can affect structures, enclosed spaces, and even partially shielded areas. The ability to use air in the explosive process also enables a different type of energy release than with solid explosives.

Fuel-air explosives can also produce a longer-lasting pressure effect. Unlike a quick, sharp explosion, the pressure wave from a fuel-air blast can last slightly longer, increasing its impact on certain targets. This characteristic has been one of the reasons they have been studied and used in specific military situations.

However, the disadvantages and dangers of fuel-air explosives are significant. One of the main concerns is the lack of precision. Because the fuel cloud spreads in the air, it can be affected by wind and other environmental factors. This means that the exact area of impact may be difficult to control. As a result, there is a higher risk of unintended damage to nearby areas.

Another major issue is the impact on people. The pressure wave and heat generated by the explosion can cause serious harm. Injuries may result from blast force, burns from the fireball, or damage from collapsing structures. The wide-area effect increases the likelihood that people who are not the intended target may be affected.

The impact on human life can be severe. In addition to immediate injuries, there can be long-term effects on survivors. The psychological impact of experiencing such a powerful and widespread explosion can also be significant. Communities may feel fear and insecurity, especially if such weapons are used repeatedly.

Human settlements can be heavily affected by fuel-air explosions. Buildings, roads, and infrastructure within the affected area may be damaged or destroyed. Essential services such as water supply, electricity, and communication can be disrupted. In urban areas, the effects can be more serious due to the density of people and structures. Recovery from such damage can take time and require significant resources.

The economic impact is also important. The financial burden of repairing or rebuilding infrastructure that has been damaged can be considerable. Businesses may suffer losses if their operations are disrupted. In addition, the development and deployment of such weapons require investment in research, materials, and technology. These costs can add to the overall economic burden.

From an environmental perspective, fuel-air explosives can cause damage to the immediate surroundings. The explosion can destroy vegetation, disturb soil, and create debris. The burning of fuel can also produce smoke and other byproducts that affect air quality. While the environmental impact may be more localized than that of some other weapons, it can still be significant in the affected area.

There are also ethical and legal considerations. The wide-area effect of fuel-air explosives raises questions about their use in populated areas. International laws aim to protect civilians and limit unnecessary harm. The use of weapons capable of affecting large areas must be carefully considered in light of these principles. Discussions continue about how to balance military objectives with the need to reduce harm to people and the environment.

Another important point is the role of technology. The development of fuel-air explosives reflects advances in chemistry, engineering, and understanding of physical processes. These advances have many positive applications in fields such as energy, industry, and research. The challenge is to ensure that such knowledge is used responsibly and not in ways that cause harm.

In conclusion, fuel-air explosives are weapons that use a cloud of fuel and air to create a powerful and wide-reaching explosion. They differ from traditional explosives in that they rely on mixing fuel with the surrounding air before ignition. While they offer certain advantages, such as a large area of effect and strong pressure waves, they also present serious risks. Their lack of precision, potential to harm civilians, and impact on infrastructure and the environment make them a cause for concern. Understanding how these weapons work and their effects helps to highlight the importance of careful consideration and responsible use in any context.

45. Incendiary Bombs

Incendiary bombs are weapons designed to start fires rather than rely only on blast force. When they are released, their main purpose is to ignite materials such as buildings, vegetation, or fuel sources, causing fires to spread across a target area. These weapons have been used in different conflicts, especially in situations where destroying infrastructure or causing disruption over a wide area was the goal. Because fire can grow and spread beyond the initial point of impact, incendiary bombs are often associated with large-scale damage.

The history of incendiary weapons goes back many centuries, when simple fire-based tools were used in warfare; however, the modern form of incendiary bombs developed during the twentieth century. During the Second World War, many countries used them in aerial attacks on cities. Large numbers of these bombs were dropped to create widespread fires, which often merged into massive firestorms. These events showed how destructive fire-based weapons could be, especially in areas with closely built structures. After the war, the use of such weapons became a subject of debate, leading to efforts to regulate or limit their use.

An incendiary bomb typically contains a material that burns at very high temperatures. This material is designed to ignite quickly and continue burning even after it lands. The bomb also includes a casing and a mechanism for releasing or dispersing the burning substance. When the bomb reaches the target, it detonates, dispersing the material, which then ignites and starts fires. Some designs focus on creating many small fires over a wide area, while others concentrate on a specific location.

The construction of incendiary bombs involves combining a flammable substance with a delivery system that ensures it spreads effectively. The materials used must ignite easily and sustain high temperatures. Engineers also design the casing to break apart in a way that helps distribute the burning material. Because of the risks and regulations associated with such weapons, detailed information about their design and production is not openly shared.

One of the main advantages of incendiary bombs, from a military perspective, is their ability to cause widespread fire damage. Fires can spread through buildings, especially in areas where structures are close together. This can make incendiary bombs effective in targeting infrastructure, storage areas, or other facilities. Another advantage is that fire can continue to cause damage even after the initial impact, increasing the overall effect.

Incendiary bombs can also be used to create confusion and disruption. Fires can block roads, damage communication systems, and make it difficult for people to respond. This can affect both immediate operations and longer-term recovery efforts. The psychological effect of fire, which can be intense and difficult to control, is another factor considered when using it.

However, the disadvantages and dangers of incendiary bombs are very serious. One of the main concerns is the lack of control over fire. Once a fire ignites, it may propagate in unforeseen manners, influenced by elements like wind, weather conditions, and the materials available in the vicinity. This makes it difficult to limit the damage to a specific target. Fires can quickly move beyond the intended area and affect nearby buildings and communities.

Another major issue is the impact on civilians. In urban areas, where people live and work close together, fires can spread rapidly and cause harm to many individuals. Burns, smoke, or collapsing structures may injure people. The risk to those who are not involved in any conflict is a key concern, and it has led to calls for restrictions on the use of such weapons.

The impact on human life can be severe. Fires caused by incendiary bombs can lead to injuries, loss of life, and long-term health effects from smoke and heat exposure. Survivors may face physical and emotional challenges as they recover. The fear and trauma associated with large fires can also affect communities for a long time.

Human settlements can be heavily affected by incendiary attacks. Buildings, homes, and public facilities may be destroyed or damaged. Essential services such as water supply, electricity, and communication can be disrupted. In some cases, entire neighborhoods may be affected, leading to displacement and loss of shelter. The process of rebuilding can take years and require significant resources.

The economic impact of incendiary bombs is also important. Damage to infrastructure and property can result in high costs for repair and reconstruction. Businesses may be disrupted or destroyed, affecting livelihoods and local economies. In addition, the cost of emergency response, including firefighting and medical care, can be substantial.

From an environmental perspective, fires can cause significant damage to natural and built environments. Vegetation may be destroyed, and wildlife can be affected. Smoke and pollutants released during fires can affect air quality and contribute to environmental problems. The long-term effects depend on the scale of the fires and the type of materials involved.

There are also ethical and legal considerations. The use of incendiary weapons has been widely debated due to their potential to cause widespread and uncontrollable harm. International laws and agreements aim to limit the use of weapons that can affect civilians or cause unnecessary suffering. These rules reflect a concern for protecting human life and reducing the impact of conflict on communities.

Another important issue is the challenge of controlling and responding to fires. Firefighting efforts can be difficult, especially when multiple fires break out simultaneously. Limited resources, damaged infrastructure, and ongoing risks can make it harder to manage the situation. This can increase the overall impact and prolong recovery.

It is also worth noting that the science of combustion and fire has many positive uses. Controlled use of fire is important in areas such as energy production, industry, and agriculture. Understanding how fire behaves helps improve safety and prevent accidents. This shows that the same knowledge can be used for both helpful and harmful purposes.

In conclusion, incendiary bombs are weapons designed to start and spread fires, causing damage over a wide area. Their use has a long history, particularly in large-scale conflicts where fire was used to disrupt and destroy. While they may offer certain advantages in terms of widespread effect, their risks and consequences are significant. The difficulty of controlling fire, the impact on civilians, and the long-term damage to communities and the environment make it a serious concern. Understanding these weapons and their effects highlights the importance of careful consideration, regulation, and efforts to reduce harm in any situation.

46. Anti-Ship Missiles

Anti-ship missiles are weapons designed to target and damage or destroy ships at sea. They are used in naval warfare and are built to travel at high speed over water toward a specific vessel. These missiles can be launched from different platforms, including ships, submarines, aircraft, or land-based systems. Their main purpose is to strike ships with accuracy and enough force to disable or sink them. Because modern ships are large, complex, and often well-defended, anti-ship missiles are developed to overcome these defenses and reach their targets.

The history of anti-ship missiles began in the mid-twentieth century, when advances in rocketry and guidance systems enabled the development of weapons capable of traveling long distances and striking moving targets. During the Second World War, early guided weapons were tested, but the technology was still limited. In the decades that followed, especially during the Cold War, many countries invested in developing more effective anti-ship missiles. These weapons became an important part of naval strategy, as they allowed forces to engage enemy ships from a distance rather than relying only on guns or torpedoes. Over time, improvements in electronics, radar, and propulsion led to more advanced and reliable systems.

An anti-ship missile generally includes several key components. It has a propulsion system that enables it to travel at high speed, often powered by a rocket or jet engine. It also has a guidance system that helps it find and follow its target. This guidance may use radar, infrared sensors, or other technologies to detect the ship and adjust the missile's path. The missile carries a warhead designed to cause damage upon impact or detonation. The missile's structure is designed to withstand the stresses of high-speed travel and maintain stability during flight.

The working of an anti-ship missile involves several stages. After launch, the missile travels toward the target area. During this phase, it may follow a programmed path or receive updates from external systems. As it gets closer, the guidance system becomes more active, searching for the target and adjusting the missile's direction. Many anti-ship missiles are designed to fly at low altitude over the sea, which helps them avoid detection by radar. When the missile reaches the target, it strikes the ship or detonates nearby, causing damage through the explosion and impact.

The development of these missiles requires advanced engineering and technology. Designers must ensure that the missile can travel long distances, maintain accuracy, and operate in challenging conditions such as weather and electronic interference. The guidance systems must be precise, and the propulsion must provide enough speed and range. Because of these complexities, governments and specialized organizations control the production and design of anti-ship missiles, and detailed technical information is not publicly shared.

One of the main advantages of anti-ship missiles is their ability to engage targets at long range. This allows forces to attack ships without coming close, reducing risk to their own vessels. Another advantage is speed. Many anti-ship missiles travel very quickly, which reduces the time available for the target to respond or defend itself. The ability to fly at low altitude also makes them harder to detect, increasing their chances of reaching the target.

Anti-ship missiles are also known for their accuracy. Modern guidance systems can track moving targets and adjust the missile's path during flight. This makes them effective against ships that are trying to maneuver or evade. Their ability to deliver a powerful warhead directly to the target can result in significant damage, even to large vessels.

However, the disadvantages and risks of anti-ship missiles are important to consider. One major concern is the potential for escalation in conflict. The use of such weapons can provoke strong responses and heighten tensions between opposing sides. Naval engagements involving missiles can quickly become more intense and widespread.

Another issue is the risk to civilian vessels. In busy maritime areas, there may be commercial ships or fishing boats present. If targeting systems are inaccurate or there is confusion, non-military vessels could be affected. This highlights the importance of careful identification and control when using such weapons.

The impact on human life can be severe. Ships often carry many people, including crew members and sometimes passengers. A missile strike can cause injuries, loss of life, and damage to the vessel. In addition to the immediate effects, survivors may face long-term physical and emotional challenges.

Human settlements can also be indirectly affected. Coastal regions may depend on shipping for trade, transportation, and the supply of goods. If maritime routes become unsafe due to the threat of anti-ship missiles, these activities can be disrupted. This can affect the availability of goods and services in nearby communities.

The economic impact of using anti-ship missiles can be significant. Damage to ships can result in financial losses for companies and industries that rely on maritime transport. Disruptions to shipping routes can affect global trade, as many goods are transported by sea. Insurance costs may increase, and businesses may face delays or additional expenses.

From an environmental perspective, damage to ships can lead to pollution, especially if the vessel is carrying fuel or other materials. Oil spills or cargo releases can affect marine ecosystems and coastal areas. The long-term effects of such incidents can be serious and may require extensive cleanup efforts.

There are also ethical and legal considerations. International law governs the conduct of warfare at sea and aims to protect civilian vessels and reduce harm. The use of anti-ship missiles must adhere to these rules, including proper target identification and efforts to minimize unintended damage. These principles are important for maintaining safety and order in maritime activities.

Another important point is the role of technology in modern naval operations. Anti-ship missiles represent a shift toward long-range and precision weapons. This has changed how naval forces operate, with a greater focus on detection, defense systems, and electronic countermeasures. Ships are now equipped with systems designed to detect and intercept incoming missiles, which adds another layer of complexity to naval engagements.

It is also worth noting that the technologies used in anti-ship missiles have applications beyond military use. Guidance systems, sensors, and propulsion technologies are used in areas such as space exploration, aviation, and communication. These technologies have contributed to many advancements that benefit society.

In conclusion, anti-ship missiles are advanced weapons designed to target and damage ships at sea. They have developed over time with improvements in speed, guidance, and accuracy. While they offer advantages such as long-range capability and precision, they also pose serious risks. Their impact on human life, maritime safety, and the environment highlights the need for careful use and strong regulation. Understanding these weapons helps to provide a clearer picture of their role in modern warfare and their broader effects on society.

47. Torpedoes

Torpedoes are weapons used underwater to aim at ships and submarines. They move through water using their own propulsion system and are guided toward a target before exploding. Unlike many other weapons that travel through the air, torpedoes operate below the surface, which makes them harder to detect. Their main purpose is to damage or sink vessels by striking them below the waterline, where ships are often more vulnerable.

The history of torpedoes goes back to the nineteenth century, when early experiments were made with underwater explosives. The first practical self-propelled torpedo was developed in the late 1800s, marking an important step in naval technology. Before this, naval battles mainly relied on cannons and direct engagement. The introduction of torpedoes changed naval warfare by allowing ships and submarines to attack from a distance and from hidden positions. During the First and Second World Wars, torpedoes were widely used and became a key part of naval strategy. Over time, improvements in propulsion, guidance, and detection systems made them more effective and reliable.

A torpedo generally consists of several main parts. It has a propulsion system, which allows it to move through water. This can be powered by engines that use fuel or other energy sources. It also has a guidance system, which helps it find and follow its target. Some torpedoes are guided by wires, while others use sensors such as sonar to locate ships or submarines. The front section contains the warhead, which is the explosive part. When the torpedo reaches its target, the warhead detonates, causing a powerful explosion. The body of the torpedo is designed to move smoothly through water and withstand pressure.

The working of a torpedo begins when it is launched from a platform such as a ship, submarine, or aircraft. Once in the water, the propulsion system starts, and the torpedo moves toward the target. The guidance system plays an important role during this phase. In some cases, the torpedo follows a preset path, while in others it actively searches for the target using sensors. Modern torpedoes can adjust their course during travel, which increases their accuracy. When the torpedo reaches close to the target, it detonates either on contact or at a short distance, depending on its design.

The development of torpedoes requires advanced engineering and careful design. Engineers must consider factors such as water resistance, pressure, and noise. A torpedo should move efficiently and quietly to avoid detection. The guidance system must be accurate, especially in environments with other sounds or obstacles. Because of the complexity and importance of these systems, their design and production are controlled, and detailed technical information is not widely shared.

One of the main advantages of torpedoes is their stealth. Since they travel underwater, they are less visible than weapons that move through the air. This makes it harder for the target to detect and respond in time. Another advantage is their effectiveness. A torpedo can cause significant damage by striking below the waterline, leading to flooding and sinking. This makes them a powerful tool in naval combat.

Torpedoes are also flexible in their use. They can be launched from different platforms, including submarines, which can operate quietly and remain hidden. This allows for surprise attacks and increases their strategic value. In addition, modern torpedoes can be designed to target different types of vessels, from large ships to smaller submarines.

However, the disadvantages and risks of torpedoes are also important to consider. One major concern is the potential for unintended targets. In busy maritime areas, there may be civilian ships or other vessels present. If identification is not accurate, there is a risk that non-military targets could be affected. This highlights the need for careful control and decision-making.

Another issue is the technology's complexity. Torpedoes rely on advanced systems for propulsion and guidance. If these systems fail or are disrupted, the torpedo may not reach its target. Environmental factors such as water conditions, temperature, and noise can also affect performance. This can reduce effectiveness and create uncertainty.

The impact on human life can be severe. Ships often carry crews, and sometimes passengers, and a torpedo strike can result in injuries or loss of life. Survivors may face physical harm as well as emotional stress. The sudden nature of underwater attacks can make them particularly dangerous, as there may be little warning.

Human settlements can be indirectly affected by the use of torpedoes. Coastal areas and communities that depend on maritime activity may experience disruption if shipping routes become unsafe. Trade and transportation can be affected, which in turn influences the availability of goods and services. This can have wider effects on daily life in these regions.

The economic impact of torpedoes can be significant. Ships represent large investments, and damage or loss can result in financial losses. Disruptions to shipping can affect global trade, as many goods are transported by sea. Insurance costs may increase, and companies may face delays or additional expenses. These factors can affect both local and global economies.

From an environmental perspective, torpedo attacks can lead to pollution. If a ship carrying fuel or cargo is damaged, substances may be released into the water. This can harm marine life and affect ecosystems. Oil spills and other forms of pollution can have long-term effects on the environment and may require extensive cleanup efforts.

There are also ethical and legal considerations. Naval warfare is governed by international law that aims to protect civilian vessels and minimize harm. The use of torpedoes must follow these rules, including proper identification of targets and efforts to avoid unnecessary damage. These principles are important for maintaining safety and responsibility in maritime operations.

Another important point is the role of torpedoes in modern naval strategy. They have influenced how ships and submarines are designed and operated. Many vessels are equipped with systems to detect and counter incoming torpedoes, which adds another layer of complexity to naval engagements. This ongoing development reflects the balance between offensive and defensive technologies.

It is also worth noting that the technologies used in torpedoes have connections to other fields. Underwater navigation, propulsion, and sensor systems are used in areas such as ocean research, exploration, and communication. These technologies have helped expand knowledge about the oceans and support various industries. This shows that the same scientific principles can be applied in both beneficial and harmful ways.

In conclusion, torpedoes are underwater weapons designed to target ships and submarines with precision and force. They have a long history and have become an important part of naval operations. While they offer advantages such as stealth and effectiveness, they also pose serious risks. Their impact on human life, maritime safety, and the environment highlights the need for careful use and strong regulation. Understanding torpedoes helps provide a clearer view of their role in modern warfare and their broader effects on society.

48. Tactical Drones (Armed UAVs)

Tactical drones, also known as armed uncrewed aerial vehicles, are aircraft that can fly without a pilot on board and are equipped with weapons. These systems are controlled remotely by operators or follow programmed instructions. They are used for both observation and attack, which makes them a flexible tool in modern military operations. Because they do not carry a human pilot, they can remain airborne for extended periods and operate in areas that may be too risky for traditional aircraft.

The development of drones began in the early twentieth century, when simple remote-controlled aircraft were first tested. At that time, the technology was limited, and these systems were mainly used for training or basic experiments. Over the decades, advances in electronics, communications, and computing have made drones more capable. By the late twentieth and early twenty-first centuries, drones were widely used for surveillance, gathering information about terrain and activity. The addition of weapons to these platforms marked a major change, enabling them to conduct targeted strikes. This development made tactical drones an important part of modern military strategy.

A tactical drone includes several key components. It has a body, or airframe, that enables it to fly and carry equipment. It uses a propulsion system, such as a propeller or jet engine, to stay airborne. The drone is equipped with sensors and cameras that provide information to the operator. These sensors can include visual cameras, infrared systems, and other devices that help detect and track targets. The communication system connects the drone to the operator, allowing control and data exchange. Armed drones also have weapons mounted on the aircraft that can be released or activated as needed.

The operation of a tactical drone begins with launch, which can take place from a runway, a vehicle, or a portable system, depending on the drone's size. Once airborne, the drone is controlled by an operator who may be located far from the area of operation. The operator uses a control system and receives live data from the drone's sensors. This allows the operator to monitor the environment, identify targets, and make decisions. If a strike is required, the operator can guide the drone into position and use its weapons. Some drones can also follow pre-set routes or use automated systems to assist with navigation.

The creation of tactical drones involves advanced engineering and technology. Designers must ensure that the drone is stable, efficient, and capable of carrying its equipment. The communication systems must be secure and reliable, as they are essential for control and data transfer. The integration of sensors and weapons requires careful planning to ensure accuracy and safety. Because of the complexity and importance of these systems, their design and production are controlled, and detailed technical information is not widely shared.

One of the main advantages of tactical drones is that they reduce risk to personnel. Since the aircraft is uncrewed, there is no pilot in danger if the drone is lost or damaged. This allows operations to take place in areas that may be too hazardous for human crews. Another advantage is endurance. Many drones can remain airborne for extended periods, providing continuous observation and enabling action when needed.

Tactical drones also offer precision. Their sensors and guidance systems allow operators to focus on specific targets. This can make them effective in situations where accuracy is important. The ability to combine surveillance and strike capabilities on a single platform enhances their usefulness, as information can be gathered and acted upon quickly.

However, the disadvantages and risks of tactical drones are significant. One major concern is accessibility. As technology becomes more available, drones can be used by a wider range of groups. This increases the risk that they may be used in harmful ways outside of controlled environments. The spread of such technology can create new challenges for security and regulation.

Another issue is the ethical concern related to remote operation. Decisions about targeting and action are made from a distance, which can raise questions about responsibility and judgment. The separation between the operator and the location of action may affect how decisions are made and understood. This has led to ongoing discussions about the use of such systems and the need for clear rules.

The impact on human life can be serious. While drones are often used for precise targeting, there is still a risk of unintended harm. People in or near the target area may be affected, and the presence of drones can create fear and stress. The constant possibility of observation and attack can influence how people live and move in affected regions.

Human settlements can also be affected by the use of tactical drones. In areas where drones are active, daily life may be disrupted. People may avoid certain locations or activities due to safety concerns. Infrastructure can be damaged during strikes, affecting services such as transportation, communication, and utilities. These effects can have both immediate and long-term consequences.

The economic impact of tactical drones includes both their cost and their effects. Developing and maintaining drone systems requires investment in technology, training, and operations. At the same time, their use can influence economic activity in affected areas. Disruptions to infrastructure and services can affect businesses and livelihoods. In addition, the spread of drone technology can lead to increased spending on defense and security measures.

From an environmental perspective, drones generally have a smaller direct impact compared to larger weapons systems. However, their use can still lead to environmental effects if strikes damage land or structures. The production and operation of drones also require resources and energy, which contribute to their overall environmental footprint.

There are also legal considerations. The use of armed drones is subject to international laws that aim to protect civilians and limit harm. These laws require careful identification of targets and consideration of potential risks. The use of drones has prompted discussions about how existing rules apply to new technologies and whether additional guidelines are needed.

Another important point is the role of technology in shaping modern conflict. Tactical drones represent a shift toward remote and technology-driven operations. This has changed how decisions are made and how actions are carried out. While this can increase efficiency, it also raises questions about accountability and control.

It is also worth noting that drone technology has many positive applications. Uncrewed aerial vehicles are used in areas such as agriculture, disaster response, environmental monitoring, and delivery services. These uses show how the same technology can provide benefits when applied responsibly.

In conclusion, tactical drones are uncrewed aircraft equipped with sensors and weapons, used for observation and targeted actions. They have developed over time with advances in technology and have become an important part of modern operations. While they offer advantages such as reduced risk to personnel, long endurance, and precision, they also present challenges related to accessibility, ethics, and impact on communities. Their influence on human life, settlements, and global systems highlights the need for careful use and strong regulation. Understanding these systems helps to provide a clearer view of their role in today's world and their broader effects on society.

49. Heavy Artillery Systems

Heavy artillery systems are large, powerful guns designed to fire shells over long distances. These weapons are used in warfare to strike targets far from the front line. Unlike small firearms, heavy artillery is not carried by individuals. It is mounted on large platforms, often supported by vehicles or fixed positions. The shells fired by these systems contain explosive material; upon reaching the target, they detonate, causing damage over a wide area.

The history of heavy artillery goes back many centuries. Early forms of artillery appeared when gunpowder was first used in weapons. Cannons became an important part of warfare in the medieval period, allowing armies to break through walls and defend strong positions. Over time, improvements in metalworking and engineering made these weapons stronger and more reliable. By the time of the First and Second World Wars, heavy artillery had become one of the most important tools in military operations. It was used to support ground forces, destroy enemy defenses, and control large areas of land. Since then, artillery systems have continued to evolve with better accuracy, longer range, and more advanced control systems.

A heavy artillery system typically includes several main components. The gun itself is a long-barreled weapon that directs the shell toward the target. A firing mechanism controls when the shell is launched. The system also includes a base or platform that provides stability during firing, as the force generated is very strong. Modern artillery often uses vehicles or tracked systems to move the gun into position. Ammunition is another key part, consisting of shells that may contain explosives or other materials designed for specific purposes.

Operating heavy artillery involves several steps. First, the target is identified, often using observation or technology such as radar or drones. The target's position is calculated, and the artillery system is adjusted to aim at the correct angle and direction. This process involves careful measurement to ensure accuracy over long distances. Once ready, the shell is loaded into the gun, and the firing mechanism is activated. The shell is propelled out of the barrel at high speed and follows a curved path through the air before reaching the target. Upon impact or at a set distance, the shell explodes, releasing its energy.

The creation of heavy artillery systems requires advanced engineering and precise manufacturing. The barrel must be strong enough to handle repeated firing without damage. The firing mechanism must be reliable and safe. The platform must provide stability and allow for accurate aiming. Modern systems also include electronic components for targeting and communication. Because of the complexity and importance of these systems, their design and production are controlled, and detailed technical information is not widely shared.

One of the main advantages of heavy artillery is its long range. It can strike targets that are far beyond the reach of smaller weapons. This allows forces to engage enemies without direct contact. Another advantage is its firepower. Artillery shells can cause significant damage to structures, vehicles, and defensive positions. The ability to fire multiple rounds in succession provides sustained pressure on a target, which can be important in certain situations.

Heavy artillery is also useful for supporting ground operations. It can be used to weaken enemy defenses before troops move forward. This can reduce resistance and make it easier to achieve objectives. The ability to adjust fire based on feedback allows for flexibility and responsiveness during operations.

However, the disadvantages and risks of heavy artillery are considerable. One of the main concerns is the lack of precision compared to some modern guided weapons. Although accuracy has improved, artillery shells can still affect areas beyond the intended target. This increases the risk of unintended damage, especially in populated areas.

Another issue is the potential for widespread destruction. The explosive shells can damage buildings, infrastructure, and natural surroundings. In urban environments, this can lead to significant harm to civilians and disruption of essential services. The sustained use of artillery can also create long-term damage to an area.

The impact on human life can be severe. Artillery strikes can cause injuries and loss of life, both directly from the explosion and indirectly from collapsing structures or fires. The psychological impact is also important. The sound and unpredictability of artillery fire can create fear and stress among people in affected areas.

Human settlements can be heavily affected by heavy artillery. Buildings may be destroyed or damaged, and infrastructure such as roads, water systems, and power supply can be disrupted. This can make it difficult for communities to function and recover. In some cases, people may need to leave their homes due to safety concerns, leading to displacement.

The economic impact of heavy artillery is also significant. Damage to infrastructure requires resources for repair and rebuilding. Businesses may be affected if facilities are damaged or if operations are disrupted. The cost of maintaining and operating artillery systems is also high, as they require equipment, ammunition, and trained personnel.

From an environmental perspective, the use of artillery can damage land and ecosystems. Explosions can disturb soil, destroy vegetation, and create debris. Repeated use in the same area can lead to long-term changes in the environment. The materials used in shells can also leave residues that affect the surroundings.

There are also ethical and legal considerations. The use of heavy artillery is governed by international laws that aim to protect civilians and limit unnecessary harm. These laws require careful targeting and consideration of the potential impact on people and infrastructure. The challenge lies in balancing military objectives with the need to reduce harm.

Another important issue is the evolution of artillery technology. Modern systems have become more advanced, with improved targeting, automation, and mobility. Some artillery systems can now use guided shells, which increase accuracy and reduce unintended effects. However, the fundamental nature of artillery as a powerful and wide-area weapon remains the same.

It is also worth noting that the principles behind artillery, such as ballistics and engineering, have applications in other fields. These include space exploration, construction, and scientific research. This shows that the knowledge used to develop such systems can also contribute to positive advancements.

In conclusion, heavy artillery systems are powerful weapons designed to fire explosive shells over long distances. They have played an important role in warfare for many centuries and continue to be used in modern operations. While they offer advantages such as range and firepower, they also present serious risks due to their potential for widespread damage. Their impact on human life, settlements, and the environment highlights the need for careful use and strong regulation. Understanding these systems helps provide a clearer view of their role and the challenges they pose to society.

50. Shockwave Weapons

Shockwave weapons are systems designed to produce powerful pressure waves that move through air, water, or solid materials. A shockwave is a sudden and intense change in pressure that travels outward from a source at high speed. In everyday life, small shockwaves can be seen in events such as thunder or a sonic boom, but in a weapon context, the energy is much stronger and focused to cause damage. These weapons do not always rely on fragments or direct impact. Instead, they use pressure to affect objects, structures, and even living beings.

The idea of using pressure waves as a form of force comes from basic physics. Scientists have long studied how energy moves through different materials. During the twentieth century, research into explosions and high-speed motion helped people understand how shockwaves behave. This knowledge was first applied to traditional explosives, in which a blast creates a pressure wave. Over time, researchers began to explore ways to generate and control these waves more precisely. This led to the concept of shockwave-based systems that could focus energy on a target without relying only on conventional explosives. Many of these ideas are still in the research or development stage.

Shockwave weapons can be created using different methods, depending on the design. Some systems use controlled explosions to generate a strong pressure wave in a specific direction. Others may use mechanical or electromagnetic systems to produce rapid changes in pressure. There are also concepts involving high-energy pulses that can cause a sudden expansion of air, generating a wave. Regardless of the method, the goal is to produce a powerful, focused pressure effect that can propagate outward and interact with a target.

The working of a shockwave weapon involves the rapid release of energy. When this energy is released, it compresses the surrounding medium, such as air, creating a wave that propagates outward. As the wave travels, it carries energy that can affect objects in its path. When it reaches a target, the sudden change in pressure can cause damage. For structures, this may mean cracking, breaking, or collapsing. In living beings, the effect can include organ damage from rapid pressure changes.

The development of these systems requires careful design and advanced technology. Engineers must understand how waves move and how to control their direction and strength. Materials must be chosen to handle high energy and stress. The system must also operate safely and reliably. Because of the complexity and potential risks, detailed technical information about these systems is not widely available and is controlled by specialized organizations.

One of the main advantages of shockwave weapons is their ability to affect a target without direct contact. The pressure wave can travel through air or other materials, allowing the system to influence areas that are not directly exposed. This can be useful in situations where a direct hit is difficult. Another advantage is that the effect can be spread across an area, allowing it to influence multiple points at once.

Shockwave weapons can also be designed to reduce the use of physical fragments. Traditional explosives often produce debris that can travel in many directions. In contrast, a controlled pressure wave may limit some of these effects. This has led to interest in using such systems in specialized scenarios that require different types of impact.

However, the disadvantages and risks of shockwave weapons are significant. One of the main concerns is the difficulty of control. Pressure waves can spread in ways that are influenced by the environment, such as buildings, terrain, and weather conditions. This makes it challenging to limit the effect to a specific target. The wave may reflect, bend, or spread, leading to unintended impact.

Another issue is the effect's invisibility. Unlike visible weapons, a shockwave cannot be easily seen as it travels. This can make it harder for people to recognize the danger and respond in time. The lack of visible signs can also create confusion and fear, especially in areas where such systems are used.

The impact on human life can be serious. Pressure waves can affect the body in ways that are not immediately visible. Rapid changes in pressure can harm internal organs, even without external injuries. This makes the effects difficult to detect and treat. In addition to physical harm, the psychological impact of experiencing such events can be significant.

Human settlements can also be affected by shockwave weapons. Buildings and infrastructure may be damaged by the pressure wave, especially if they are not designed to withstand such forces. Windows can shatter, walls can crack, and structures can become unstable. This can disrupt daily life and require repairs or rebuilding. The indirect effects on services such as communication and transportation can also be important.

The economic impact of these weapons can be considerable. Damage to buildings and infrastructure can result in high repair costs. Businesses may be affected if operations are disrupted or if facilities are damaged. The development and maintenance of shockwave systems also require investment in research and technology. These factors can influence both local and broader economic conditions.

From an environmental perspective, shockwave weapons can cause physical damage to land and structures, but they may not generate as much long-term pollution as some other weapons. However, the destruction caused by pressure waves can still affect ecosystems, especially if natural habitats are disturbed. The production and testing of such systems also have environmental considerations related to energy use and materials.

There are also ethical and legal considerations. The use of weapons that can cause harm without visible signs raises questions about responsibility and control. International laws aim to protect civilians and limit unnecessary suffering. The development of new types of weapons, including those based on shockwaves, requires careful consideration of these principles. Discussions continue about how to regulate such technologies and ensure they are not misused.

Another important point is the role of research in this field. The study of shockwaves has many beneficial applications, such as in medicine and engineering. For example, controlled shockwaves are used in medical treatments to break down kidney stones. They are also used in industrial processes and scientific research. This shows that the same principles can be used for positive purposes when applied responsibly.

It is also worth noting that shockwave weapons are still largely in development. While some concepts have been tested, many systems are not yet widely used. This means that their full capabilities and limitations are not completely understood. Continued research and discussion are needed to understand their potential impact and to guide their use.

In conclusion, shockwave weapons are systems that use powerful pressure waves to affect targets. They are based on scientific principles of energy and motion and represent an area of ongoing research. While they offer certain advantages, such as the ability to influence targets without direct contact, they also present serious risks. Their invisibility, potential for widespread impact, and effects on human life and infrastructure make them a cause for concern. Understanding these systems helps highlight the importance of responsible development and the need to consider their broader impact on humanity and society.

51. Psychological Warfare Tools

Psychological warfare tools are methods used to influence how people think, feel, and behave during times of conflict or competition. Unlike physical weapons, these tools do not rely on force or destruction. Instead, they aim to shape opinions, create confusion, weaken morale, or influence decisions. They can be used against soldiers, leaders, or entire populations. Because they affect the human mind, their impact can be wide and long-lasting, even without any physical damage.

The history of psychological warfare is very old. In ancient times, leaders used messages, symbols, and rumors to influence enemies and gain an advantage. For example, armies sometimes spread fear by exaggerating their strength or by using signs to create the impression of power. As communication methods improved, so did these techniques. During major wars in the twentieth century, especially the First and Second World Wars, psychological warfare became more organized. Governments used radio broadcasts, posters, and printed materials to influence both their own people and their opponents. In modern times, the rise of digital technology and the internet has made these methods faster, wider, and more complex.

Psychological warfare tools include a range of techniques. One common method is propaganda, which involves spreading information designed to influence opinions. This information may be true, false, or a mixture of both. Another method is misinformation, or disinformation, in which incorrect or misleading information is shared to create confusion or distrust. These messages can be spread through media, social networks, or other communication channels. Other tools include intimidation, symbolic actions, and controlled messaging that aim to shape how people understand events.

The way these tools are used often involves careful planning and communication. Messages are created with specific goals in mind, such as reducing confidence, increasing fear, or influencing behavior. These messages are then delivered through channels that reach the intended audience. In modern settings, digital platforms play a major role by allowing information to spread quickly across large populations. The process does not involve physical construction like traditional weapons, but it does require knowledge of communication, psychology, and technology.

One of the main advantages of psychological warfare tools is their ability to influence large numbers of people without direct physical action. A single message can reach many individuals and shape how they think or act. This can reduce the need for physical conflict in some situations, as changes in perception or morale can affect outcomes. Another advantage is flexibility. Messages can be adjusted based on the situation, and different approaches can be used for different audiences. Psychological tools can also operate over long distances and across borders. With modern communication systems, information can travel almost instantly. This allows influence to be exerted in areas that may be far from the source. In addition, these tools can be used continuously, with messages repeated and reinforced over time.

However, the disadvantages and risks of psychological warfare are serious. A big worry is the distribution of incorrect or deceptive information. When people receive incorrect information, it can lead to poor decisions, confusion, and loss of trust. This can affect not only individuals but also institutions and systems that rely on accurate information.

Another issue is the difficulty of control. Once information is released, it can spread in unexpected ways and reach unintended audiences. Messages can be misunderstood or altered as they are shared, leading to outcomes that differ from what was intended. This unpredictability makes it hard to manage the full impact.

The effect on human life can be significant, even without direct physical harm. Psychological warfare can create fear, stress, and uncertainty. People may feel anxious or unsure about what is true, which can affect their mental well-being. In some cases, this can lead to panic or harmful behavior, especially if people act on false information. Human settlements can also be affected. Communities may experience division if different groups believe different versions of events. Trust between individuals and institutions can weaken, making it harder to work together and solve problems. In extreme cases, misinformation can lead to unrest or conflict within a society.

The economic impact of psychological warfare can be wide-ranging. If misinformation affects markets, businesses, or public confidence, it can lead to financial losses. For example, false information about a situation can cause people to change their behavior, which may disrupt trade or services. The cost of addressing and correcting misinformation can also be high, as it requires effort from governments, organizations, and communities.

From an environmental perspective, psychological warfare does not directly damage the physical environment. However, its indirect effects can influence how resources are used and managed. For example, panic or confusion may lead to changes in consumption or behavior that have environmental consequences. The broader impact depends on how people respond to the information they receive.

There are also ethical and legal considerations. The use of psychological methods to influence people raises questions about truth, responsibility, and fairness. International laws and guidelines aim to protect individuals and societies from harmful manipulation, but the rapid spread of digital communication makes it challenging to enforce these rules. The balance between free expression and the prevention of harmful misinformation is an ongoing issue.

Another important point is the role of technology. Modern communication systems, including social media and digital platforms, have increased the reach and speed of psychological tools. This has made it easier to share information, but also easier to spread misinformation. The design of these systems can influence how messages are received and shared, which adds another layer of complexity.

It is also worth noting that the same tools used in psychological warfare can have positive uses. Communication and information sharing are essential for education, awareness, and cooperation. When used responsibly, these tools can help inform people, support decision-making, and strengthen communities. The challenge is to ensure they are used in ways that promote truth and understanding rather than harm.

Another factor to consider is resilience. Societies with strong systems for education, critical thinking, and communication are better able to resist harmful influences. Awareness of how information can be used and misused is an important step in reducing the impact of psychological warfare. Efforts to improve media literacy and transparency can help people make informed decisions.

In conclusion, psychological warfare tools are methods that use information and communication to influence people and shape outcomes. They have a long history and have evolved with advances in technology. While they offer advantages such as a wide reach and flexibility, they also pose serious risks. The spread of misinformation, the impact on mental well-being, and the potential for social division make them a significant concern. Their effects on human life, communities, and systems highlight the importance of responsible use, strong regulation, and public awareness. Understanding these tools helps provide a clearer view of how information can influence society and why it must be handled with care.

***

****

www.ingramcontent.com/pod-product-compliance
Lightning Source LLC
LaVergne TN
LVHW090358210926
848480LV00002B/377